SCREWCUTTING IN THE LATHE

SCREWCUTTING IN THE LATHE

Martin Cleeve

Special Interest Model Books

Special Interest Model Books Ltd.
Stanley House
3 Fleets Lane
Poole
Dorset
BH15 3AJ

First published by Argus Books Ltd. 1984

Reprinted 1986, 1989, 1990, 1992, 1995, 1997, 1999, 2002, 2003

This edition published by Special Interest Model Books Ltd. 2002

Reprinted 2003, 2005, 2006

© Special Interest Model Books Ltd. 2006

The right of Martin Cleeve to be identified as the Author of this work has been
asserted by him in accordance with the Copyright, Designs and Patents
Rights Act of 1988.

All rights reserved. No part of this book may be reproduced in any form by
print, photography, microfilm or any other means without written permission
from the publisher.

ISBN 0-85242-838-2

ISBN-13: 978-0-85242-838-2

Printed and bound in Malta by Progress Press Co. Ltd.

CONTENTS

ACKNOWLEDGEMENT

I am greatly indebted to the Editor of MACHINERY'S SCREW THREAD BOOK (Ed 20) for his kind permission to make use of information contained therein. Indeed without such guidance it would have been impossible to make any sound pronouncements on thread depths, basic sizes, and thread gauging methods. However, apart from space considerations, it would obviously be unfair to reproduce large verbatim extracts from the SCREW THREAD BOOK, so for those requiring more detailed information on threads, as distinct from producing them, I can but recommend the SCREW THREAD BOOK itself.

Martin Cleeve

Publisher's Note

The publishers regret to record the death of the author, after submitting his manuscript but before it had been typeset.
'Martin Cleeve' was a pen-name used by Kenneth C. Hart, a respected con-tributor for some 30 years to the Model Engineer. *His painstaking, perfectionist approach to high-quality, accurate work, which so clearly comes through in this book as in all his other writing, led him to design and describe many original lathe accessories which have been made and are regularly used in hundreds of amateur and professional workshops alike, perpetuating the memory of an engineer for whom only the highest standards would suffice.*

SECTION 1

Introduction

It has been said that lathe screwcutting cannot be taught from books, which seems to imply that students must learn this particular skill from trial and error after gathering a few basic facts from an instructor. However, this outlook may arise partly from the fact that few general engineering books can spare the necessary space, and partly because writers seldom take the trouble to make any specialised study of lathe screwcutting, with the result that the same few scraps of information are handed down from generation to generation without any attempt at sorting the wheat from the chaff; perhaps to disguise this deficiency it is sometimes remarked that too much emphasis can be placed upon the ability to cut threads in lathes. However, in this respect, while ordinary turning calls for the use of little more than common sense, efficient and time-saving lathe screwcutting cannot be undertaken on the same basis, and if a lathe operator is not in possession of all the relevant facts he may not be able to avoid wasting time: time which on small batch production can sometimes amount to whole working weeks, not just the odd 30 minutes. For example, it is not always necessary to follow the time-wasting instruction: 'For all other threads, reverse the lathe' (an instruction referring to tool repositioning between threading passes). Moreover, the adverse conditions for which lathe reversal is supposed always to be necessary can sometimes be turned to advantage for indexing the starts of multiple-start threads by a method whereby, after an initial setting, indexing takes place between every single threading pass without additional attention from the operator, and having the advantage that all starts (individual helices) are machined to identical proportions to close limits.

Having said that, it would only be fair to add that on deciding it might be a good idea to commit to paper the results of my researches, I had no idea that the describing of what is basically a simple process would call for such a plethora of writing, (and I have not used two words where one will serve) or indeed that the project would lead to two Patent Applications, one for an independently retractable and swing lathe toolholder (No. 1335978 — now lapsed), and one for a simple thread tool sharpening jig (No. 1417351 — not 'Sealed' although printed by the Patent Office), or that I would be devising formulas for the design of leadscrews of special lead for the automatic indexing of the starts of multiple-start threads when these cannot be auto-indexed from standard English or metric leadscrews.

In general, despite the rapid advancement in fully automatic machine control, the ordinary centre lathe is likely to remain with us for a long time for the reason that it does not pay to set an automatic machine for only one or a few threaded components such as those required for jig and tool-making, or for experimental and prototype work. And in many instances, even when the quantity of components reaches the 50 to 150 total, a centre lathe can offer a saving when compared with the cost of a more specialised machine and the time taken to set it.

On the other hand, automatic and semi-automatic threading attachments can now be obtained for use with standard centre lathes, and such attachments can be fairly quickly set. However, the initial cost can be high, and this has to be weighed against the quantity of threading likely to be called for.

In contrast to the foregoing, I have heard it remarked that screwcutting facilities are not really necessary on centre lathes these days, as all threads can be cut with taps and dies. Now although modern taps and die-heads are capable of cutting clean bright threads to close limits, their use sometimes calls for very high torques, whereas a centre lathe always forms threads in easy stages, admirably suited to those components which by nature of their design could not be gripped with sufficient security to withstand the high torques imposed when tap or die running. Moreover a lathe will cut a thread of *any* pitch on *any* diameter: for example it is as easy to cut 16 tpi on a diameter of 4 in. as on a diameter of $\frac{1}{2}$ in. or less, whereas the use of taps and dies limits one to standard sizes, and when only a few special threads are called for one obviously would not wish either to pay the high cost of special taps or dies, or to await delivery when such threads can be lathe screwcut for the trifling cost of a single-point threading tool and a few minutes of a lathe operator's time. Similar remarks of course apply if a standard size tap or die is not in stock.

There is also the point that bores to be threaded are sometimes very short or shallow, a total depth being limited to say $\frac{3}{16}$ in. or so (4.8 mm) with an abrupt shoulder or completely closed base. These threads are impossible to cut with a tap simply because the tap would 'bottom' before the necessary tapered lead had fully entered, whereas such threads are easily lathe screwcut with a single-point tool. I have also encountered external threads that were required to continue inside a recess – where of course no die could operate, and these had to be cut by the use of a special cranked threading tool. Another point in favour of lathe screwcutting is that threads so produced are concentric and symmetrically disposed about a component axis to close limits – i.e. are 'square' to axis.

METRICATION

Those brought up entirely with metric units will have no difficulty in following the recommendation that, with metrication, designers and engineers should work entirely from metric concepts. However, those of us long accustomed to working to English imperial measure tend to feel uncomfortable until we have converted metric figures into English units having a satisfactory meaning to us. For example, for a time we will not have a clear idea of the implication of a thread pitch error of, say, minus 0.003 mm until we have converted to inch measure and found that 0.003 mm equals 0.000118 in., or just over 1/10 thou/inch. In this respect, too, many centre lathes will probably remain in use with English feed dials graduated in thousandths of an inch, and metric thread sizing will have to be carried out to inch standards. The object here therefore is to deal with these problems of change in such a way that the reader may choose a line of action best suited to his particular need, and simple formulas are given to facilitate working to either metric or English units. As a matter of fact, partial metrication has led to the writer often having to lathe screwcut batches of 50 or 100 components with an English thread at one end, and a metric thread at the other end.

CONVERSIONS

Fortunately these days it is possible to buy a good basic electronic calculator for a very modest sum, so it is no longer necessary to occupy valuable space with conversion tables. Indeed, with a basic formula and a calculator, any necessary figures can be obtained far more pleasantly, quickly and accurately than by thumbing through fully tabulated data.

GENERAL FORMULAS

The following formulas will be useful for general reference:

1 To convert inches to millimetres, multiply inches by 25.4.

2 To convert millimetres to inches, multiply by 0.03937, or divide by 25.4.

3 Given the pitch of a thread in millimetres, find the threads/inch:

$$\text{Threads/inch} = \frac{25.4}{\text{Pitch in mm.}}$$

4 Given the threads/inch, find the pitch in mm:

$$\text{Metric pitch (mm)} = \frac{25.4}{\text{Threads/inch}}$$

5 Given the inch pitch, find the metric pitch in mm:

$$\text{Metric pitch (mm)} = \text{Inch pitch} \times 25.4$$

6 Given the pitch in millimetres, find the inch pitch:

Inch pitch =
0.03937 x. Metric pitch in mm

or

$$\text{Inch pitch} = \frac{\text{Metric pitch}}{25.4}$$

7 Given the threads/inch, find the pitch in inches:

$$\text{Inch pitch} = \frac{1}{\text{Threads/inch}}$$

8 Given the pitch by inch measure, find the threads/inch:

$$\text{Threads/inch} = \frac{1}{\text{Inch pitch}}$$

9 Given the metric pitch (mm), find the threads per centimetre:

$$\text{Threads/cm} = \frac{10}{\text{Pitch in mm}}$$

10 Given the threads/inch, find the threads/cm:

$$\text{Threads/cm} = \frac{\text{Threads/inch}}{2.54}$$

NOTE: The notation 'threads/centimetre' is not ordinarily used or recognised, but is sometimes useful for explanatory purposes associated with lathe leadscrew gearing.

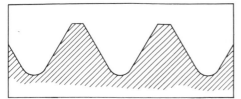

The International Standardisation Organisation (ISO) metric screw thread form.
60 deg. included thread angle.
Screw thread crests may be rounded inside the maximum outline: rounding is optional.
Root radius = 0.1443 × Pitch. (Also optional)

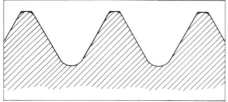

Unified & American screw thread form.
60 deg. included thread angle.
Thread crest may be flat, or given a radius of 0.108253 × Pitch.
Root radius =0.144338× Pitch. (Also optional)

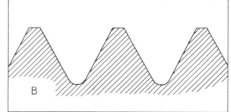

The Whitworth & British Standard Fine (BSF) screw thread form.
55 deg. included thread angle.
Crest and Root radius = 0.1373292 × Pitch
A The true form. B as lathe screwcut with a single-point tool.

QUICK REFERENCE THREAD INFORMATION SUMMARY
DEPTH OF THREAD. (SCREW). BASIC DESIGN DEPTH

ISO Metric 60 deg.

By mm

D = Pitch (mm) × 0.6134

By inch measure:

D = Pitch (mm) × 0.0241*

* This figure is a close approximation.

UNIFIED 60 deg.

By inch measure

$$D = \frac{0.6134}{\text{Threads/inch}} \quad \text{or}$$

D = 0.6134 × Pitch (inch)

By mm:

$$D = \frac{15.58}{\text{Threads/inch}} \quad \text{or}$$

D = Pitch (inch) × 15.58

WHITWORTH & BSF 55 DEG.

By inch measure:

$$D = \frac{0.64}{\text{Threads/inch}} \quad \text{or}$$

D = Pitch × 0.64

WHITWORTH & BSF 55 DEG.

By mm:

$$D = \frac{16.256}{\text{Threads/inch}} \quad \text{or}$$

D = Pitch (inch) × 16.256

NUT BORE (MINOR DIAMETER) SIZING. RECOMMENDED MINIMUM

ISO Metric. 60 deg. By inch measure.
BORE = Major nominal screw dia (by inch measure) minus (Pitch (mm) × 0.0426)

ISO Metric, 60 deg. By millimetres.
BORE = Major nominal screw dia (mm) minus (Pitch x 1.0825)

UNIFIED 60 deg. By inch measure.

$$\text{BORE} = \text{Major nominal screw dia. minus} \left(\frac{1.0825}{\text{Threads/inch}}\right)$$

UNIFIED 60 deg. By millimetres.

$$\text{BORE} = \text{Major nominal screw dia. (mm) minus}\left(\frac{27.5}{\text{Threads/inch}}\right)$$

WHITWORTH AND BRITISH STANDARD FINE 55 deg.
By inch measure.

$$\text{BORE} = \text{Major nominal screw dia. minus}\left(\frac{1.2}{\text{Threads/inch}}\right)$$

WHITWORTH AND BRITISH STANDARD FINE 55 deg.
By millimetres.

$$\text{BORE} = \text{Major nominal screw dia. (mm) minus}\left(\frac{30.48}{\text{Threads/inch}}\right)$$

NUT BORE SIZING BY PERCENTAGE OF FULL THREAD

$$\text{BORE} = \text{Major nominal screw dia. minus}\left(\frac{2d \times \% \text{ required}}{100}\right)$$

where d = standard basic depth of corresponding SCREW thread. % required = percentage of full thread engagement required.

NUT THREAD DEPTHS
(Nut thread depths are taken from the surface of bores slightly larger than would be given by major screw diameter minus twice the depth of thread of the corresponding screw, hence basic nut thread depths are less than corresponding screw thread depths, and are really only useful as a guide. Actual nut thread depths may be greater or less than calculated).

ISO Metric. 60 deg. Depth of NUT thread by mm:

$$D = \text{Pitch (mm)} \times 0.5418$$

Depth of NUT thread by inch measure:

$$D = \text{Pitch (mm)} \times 0.0213$$

UNIFIED. Depth of NUT thread by inch measure:

$$D = \frac{0.5418}{\text{Threads/inch}}$$

Depth of NUT thread by millimetres:

$$D = \frac{13.76}{\text{Threads/inch}}$$

WHITWORTH AND BRITISH STANDARD FINE

Depth of NUT thread by inch measure:

$$D = \frac{0.6}{\text{Threads/inch}}$$

Depth of NUT thread by millimetres:

$$D = \frac{15.24}{\text{Threads/inch}}$$

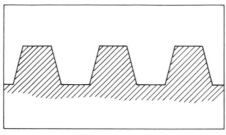

*The Acme screw thread form.
29 deg. included thread angle.*

THE ACME FORM THREAD 29 deg.

DEPTH OF THREAD – SCREW

By inch measure:

$$D = \frac{1}{2 \times \text{Threads/inch}}$$

By millimetres:

$$D = \frac{12.7}{\text{Threads/Inch}}$$

BASIC DESIGN DEPTH

plus 0.010

plus 0.254

NUT BORE (MINOR DIAMETER) SIZING

BORE = Major nominal screw diameter minus pitch.

(Nut thread depth is the same as screw thread depth)

NOTE: For the Acme thread (and for the trapezoidal form) the standard clearances between screw and nut appear to be extraordinarily liberal. Taking as an example a thread of $\frac{5}{8}$ in. dia. × 8 threads/inch, the screw-thread depth is 0.0725 in. leaving a root diameter of 0.480 in., yet the recommended nut bore is 0.500 in., showing that a screw thread depth of about 0.064 in. (1.63 mm) would be sufficient, unless, of course, contrary instructions are received. Similarly, the major diameter of a $\frac{5}{8}$ in. dia × 8 threads/inch ground thread tap is 0.654 in., i.e. 0.029 in. in excess of major screw diameter, thus offering an 'annular' thread clearance of 14.5 thou./inch (0.37 mm) which, to say the least, appears to offer a somewhat excessive space 'for lubrication', especially when compared with the much smaller clearances recommended for plain shafts and bearings.

THE TRAPEZOIDAL METRIC THREAD   30 deg.   (Similar to the Acme form)

DEPTH OF THREAD – SCREW   BASIC   DESIGN DEPTHS.

(Thread depths are not proportionate to pitch)

PITCH mm	DEPTH OF SCREW THREAD mm	 Inch			
			6.0	3.25	0.1279
			7.0	3.75	0.1476
2.0	1.25	0.0492	8.0	4.25	0.1673
3.0	1.75	0.0689	9.0	4.75	0.1870
4.0	2.25	0.0886	10.0	5.25	0.2067
5.0	2.75	0.1083	12.0	6.25	0.2461

NUT BORE (MINOR DIAMETER) SIZING

For nut bores the most practical approach appears to lie in use of the percentage-of-full-thread formula, unless instructed otherwise.

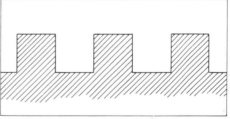

The Square thread screw form.

THE SQUARE THREAD FORM.
Thread flank angle: 90 deg.

DEPTH OF THREAD: SCREW –
By English or metric measure:
$$D = 0.5 \times \text{Pitch}$$

WIDTH OF THREAD SPACE –
(Screw) W = 0.5 × Pitch.

NUT BORE SIZING (Minor diameter)
By English or metric measure:
Bore = (Major screw dia. minus Pitch) plus *C*

where C = a clearance allowance varying with Pitch.

(Without a "clearance allowance" the crests of a nut thread would contact or interfere with the root of a correspondingly basic sized square thread screw)

NUT THREAD DEPTH
As sized from the inner surface of (a slightly enlarged) minor nut diameter, nut thread depth will be the same as the screw thread depth.

The clearance allowance may be any amount felt desirable for lubrication, unless of course, precise instructions are given.

For side (flank) clearance, the thickness of the body of a nut thread will also be slightly less than the 0.5 × P. space dimension of the corresponding screw thread.

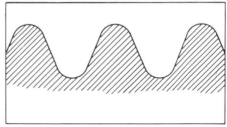

The British Association (BA) screw thread form.
47½ deg. included thread angle.
Radius at Crest and Root = 0.1808346 × Pitch.
Depth of thread = 0.6 × Pitch.

15

The Principles of Lathe Screwcutting

The drawing, Fig. 1, shows in an elementary way the principles of thread cutting by means of a master screw: a leadscrew (pronounced 'leed', by the way). Points to note are that the spindle, which is revolving with the chuck and component to be threaded, drives the leadscrew through gearing: in this example by two gears each having 45 teeth and therefore giving a ratio of 1:1. By this means the leadscrew will revolve at exactly the same speed as the piece to be screwed, and at the same time will cause the nut (which is prevented from rotating) to move from right to left by a certain distance for each revolution of the leadscrew. If the leadscrew has 8 threads to the inch, or a pitch of $\frac{1}{8}$ inch, each exact revolution of the leadscrew will cause the nut to advance $\frac{1}{8}$ inch. If the nut is made to carry a suitable holder provided with a pointed tool, and this is brought into contact with the truly cylindrical workpiece, then a helix will be circumscribed thereon, and the distance between any two adjacent helices will be $\frac{1}{8}$ in., quite regardless of the actual diameter of the workpiece and regardless of the actual speed of rotation, because if the work

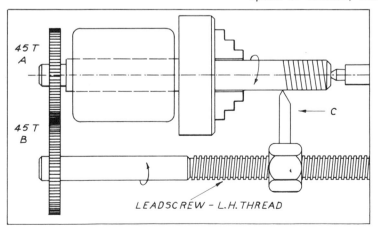

45 T
A

45 T
B

C

LEADSCREW – L.H. THREAD

Fig. 1 Illustrating the basic principles of lathe screwcutting.

Fig. 2. Inside view of the carriage apron of a small lathe. The pinion at the left engages with a rack fixed to the lathe bed. The half-nuts may be seen at the right, and leadscrew indicator is fitted at the left.

(The plummer-block type bearer held a non-standard anti leadscrew deflection bush. This became unnecessary with a change to the square thread form leadscrew.)

speed is altered, so is the leadscrew speed in the same proportion.

In practice the nut is split into two pieces or "halves" each provided with a slideway backing, mounted in corresponding guideways so that by means of a hand-lever and cam-type mechanism each half can be moved radially outwards, thus disengaging the leadscrew. The leadscrew nut thus becomes known as "the half-nuts", "the clasp nut", or the "split nut".

The photograph Fig. 2 is an inside view of the apron of a small lathe and will give an idea of the arrangement. The half-nuts are shown in the disengaged position. The small pinion at the left engages with a rack for hand traversing the lathe carriage when required.

A pair of half-nuts suitable for the apron shown may be seen in the photograph, Fig. 3.

Referring again to our basic diagram, Fig. 1, the initial helix circumscribed on the workpiece may be regarded as the first of a series of "cuts" or "threading

passes" as may be seen again at the foot of Fig. 4 which, if read upwards, shows how a screw thread is formed by a succession of passes each a little deeper than the previous one, until the thread is complete. The diagram, of course, indicates only a few of the greater number of passes required before a full depthing and sizing is reached.

Fig. 3. A pair of half-nuts for use in a small lathe.

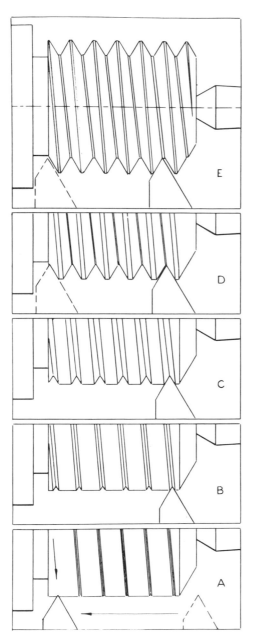

Fig. 4. Showing how a screw thread is formed by a succession of cutting passes of progressively increasing depth.

ALTERING THE PITCH. CALCULATIONS

From what has already been said it follows that if the leadscrew (Fig. 1) can be caused to revolve at exactly one half the speed of the component, and the leadscrew has 8 threads to the inch, then for each half revolution of the leadscrew the component will make one complete turn and one complete helix will be circumscribed. One complete helix for each half revolution of the leadscrew equals 16 complete helices for 8 revolutions of the leadscrew. For each 8 revolutions of the leadscrew the tool will move through a distance of one inch: accordingly 16 helices or threads to the inch would be formed on the component.

In our basic example (Fig. 1) the leadscrew could be made to rotate at half the speed of the component by removing the two 45 teeth gears, A and B, and fitting a driver of 30 teeth at A, and a driven of 60 teeth at B, on the leadscrew.

Actually, of course, it is not possible to so relate the distance between the lathe spindle and the leadscrew that no more than two gears of equal or different size may be arranged to meet all ratio needs, so what is known as a "quadrant" or "change gear arm" is provided, upon which intermediate gearing may be assembled and adjusted not only for desired ratios, but to bridge the gap between the lathe spindle or tumbler reverse and the leadscrew gear.

The photograph Fig. 5 shows a typical arrangement for a small lathe of the instrument type. Each of the slotted quadrant arms carries a movable "stud" for the intermediate gearing, and the whole quadrant may be pivoted about the

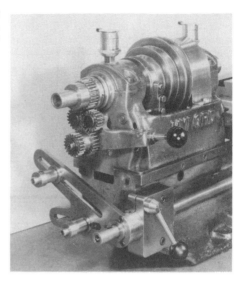

leadscrew axis by releasing the locking handlever. This illustration also shows a tumbler reverse mechanism which may be seen in its three positions in the diagram Fig. 6.

Some earlier lathes of this kind were sold without a tumbler reverse mechanism, but when one is fitted, suitable driving wheels for the quadrant gearing are mounted on an extension spigot S which is integral with the final driven gear G of the tumbler reverse. For later explanations it will be convenient to refer to gears fitted to this spigot as "first gear drivers" and to call the spigot itself "the tumbler reverse output spigot".

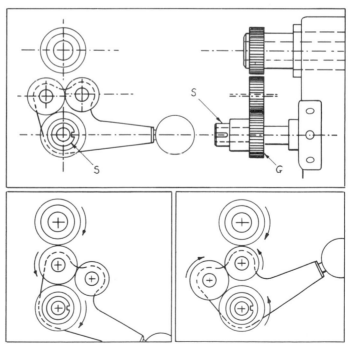

Fig. 6. A tumbler-reverse mechanism shown in three positions: neutral above, forward and reverse below.

19

Normally on lathes of this kind, the first gear driver will rotate at exactly the same speed as the lathe spindle. The tumbler reverse is used either to cause the leadscrew to revolve "backwards" for cutting left-hand threads, or to correct the direction of rotation of the leadscrew in the event of a gear train being of a nature that makes a correction necessary.

For a simple lathe of the type illustrated, a set of gears is provided, and with them it is possible to assemble a great variety of ratios between the lathe spindle and the leadscrew. These gears are known as "change gears". Special mention is made of these because of certain differences in the way in which sets are sometimes made up. For this particular lathe it is customary to provide a set of gears as follows: Two having 20 teeth, and one each of 25, 30, 35 and so on up to 75 teeth together with one of 38 teeth for reasons which will be explained later. However, in future such a set will be referred to as "20-75 by fives" or merely as a "set rising by fives". Change gears rising in size by four teeth at a time, say 24, 28, 32 and so on are not unknown and, of course, such a set would be referred to as "rising by fours". But what should be noted is that for example, a gear of 32 teeth would be "special" to a set rising by fives while a 55 teeth gear would be "special" to a set rising by fours and a gear of 33 teeth would be "special" to both sets.

Before giving a general formula for change gear calculations it will be helpful to consider the basic requirements for cutting a thread of 24 to the inch with a leadscrew of 8 threads to the inch.

If we wish to cut, over a given length, three times as many helices on a component as are contained in the same length of leadscrew, the leadscrew must rotate at one third the speed of the component.

This can be arranged by using a 20 teeth gear as a first gear driver and a 60 teeth gear on the leadscrew, but as these two gears will be positioned too far apart for direct meshing, the gap is bridged with spare change gears, which for this purpose become temporarily known as "idle gears", or "idlers". Any number of idle gears may be interposed without affecting the ratio between the first driver and the last given gear although design limitations usually restrict the possible number of idlers to two. The diagrams, Fig. 7, will give an idea of the necessary 1:3 ratio, the left hand drawing showing the gearing "straightened out" for clarity, and the right hand drawing showing the gearing as it would be assembled on the lathe.

The idlers A and B, Fig. 7, are shown as a 65 and 40, but their actual size is of no importance provided they are capable of bridging the space between the first 20 driver and the last 60 driven.

Some find it difficult to understand that the interposition of one or more idle gears cannot affect the ratio between the first driver and the last driven gear. One way of looking at the question is to consider that the teeth velocities of the intermediate idlers must be exactly the same as the teeth velocity of the first driver, therefore the effect of meshing the leadscrew gear with the idler gear cannot differ from the effect of meshing the leadscrew gear directly with the first gear driver. Again, idle gears can no more affect the ratio between the first driver and the last driven than could a chain, or the number of links in a chain that may be needed to couple the first and last gear. What does happen is that small idlers will revolve more quickly and large idlers more slowly relative to the first driver, but as a drive is not being taken from the *hub* of the idle gear, or gears, their speed of rotation can

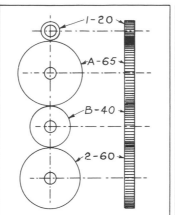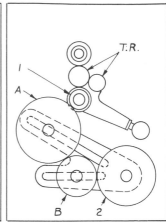

Fig. 7. An example of a simple gear train. Gearing shown is for 24 tpi with an 8 tpi leadscrew.

At the left the gear train has been 'straightened out' for clarity. At the right the same gearing is shown as assembled on a quadrant. The tumbler-reverse (TR) although shown, plays no part in the ratio.

be of no consequence. It is worth noting, however that in a manner similar to that of the tumbler reverse, with the interposition of one idle gear the direction of rotation of the last driven gear will be the same as that of the driver, and the interposition of two idle gears will reverse the direction of rotation of the final driven gear relative to the first driver.

That the leadscrews of some of the smaller lathes have left hand threads may be explained by the fact that a handwheel, which can be fitted to the leadscrew at the right-hand end, may be turned clockwise to feed the carriage towards the chuck.

Before continuing with details of a general formula, it will be convenient to mention that although a leadscrew of 8 threads to the inch appears to be the standard today for the range of smaller lathes, an earlier machine may be found to have a leadscrew of 10 threads/inch. With larger industrial lathes having leadscrews of 4 or even 2 threads/inch we are not really concerned at this stage because they will be fitted with selective gearboxes and calculations will not normally be required. Accordingly, to keep explana-

tions within reasonable bounds it was felt best to deal chiefly with calculations for leadscrews of 8 threads/inch. Metric leadscrews will be dealt with later.

The simple examples already given for 8, 16 and 24 tpi with a leadscrew of 8 tpi showed that gearing was required in the ratios 1:1, 1:2 and 1:3, or, in terms of the number of threads to the inch of the leadscrew to the number of threads to the inch for which the lathe was to be geared, 8:8, 8:16, and 8:24, and finally in terms of the number of teeth in the driving and driven gears: 45:45 (or any two of equal size), 30:60, and 20:60.

Accordingly, the number of teeth in the driving gear divided by the number of teeth in the driven gear, or leadscrew gear, is equal to the number of threads to the inch of the leadscrew divided by the number of threads to the inch for which the lathe is to be geared, a statement which may be condensed to the convenient form:

$$\frac{\text{Drivers}}{\text{Driven}} = \frac{\text{Threads/inch of leadscrew}}{\text{Threads/inch required}}$$

The abbreviated form Drivers/Driven will be used in all subsequent examples.

21

The reason for the plural in Drivers arises from the fact that there may be more than one driver and more than one driven gear in a "compound train", as will be explained shortly. But when there are more than one of each, the expression Drivers/Driven should be read as "The multiple of the number of teeth in individual driving gears divided by the multiple of the number of teeth in individual driven gears."

USE OF FORMULA

Suppose we wish to gear a lathe for a thread of 9 to the inch, and the leadscrew is of 8 to the inch, substituting the known figures we have:

$$\frac{\text{Drivers}}{\text{Driven}} = \frac{8}{9}$$

accordingly a first driver of 8 teeth driving a leadscrew wheel of 9 teeth would give the desired ratio, but as gears of only 8 and 9 teeth would be impracticable we have to multiply both numerator (8) and denominator (9) by some number that will increase the number of teeth to a convenient figure. If it is known that the change gears rise in sizes by five teeth increments, then there is no point in multiplying both numerator and denominator by any number except 5, or multiples of 5:

$$\frac{\text{Driver}}{\text{Driven}} = \frac{8 \times 5}{9 \times 5} = \frac{40}{45}$$

Had the gears risen by increments of four teeth, the gearing, with an 8 t.p.i. leadscrew would become:

$$\frac{\text{Driver}}{\text{Driven}} = \frac{8 \times 4}{9 \times 4} = \frac{32}{36}$$

In either instance, of course, it would be necessary to interpose one or two idle gears of any convenient size to bridge the gap between the first driver and the leadscrew gear. One is often sufficient, although it suited the writer's purpose to keep two gears spare to the set (a 65 and 40) more generally in place on the intermediate quadrant studs. But this means a larger range of screw gearing can be set by changing only the one leadscrew gear and moving the idlers to suit the new diameter.

Further examples similar to the foregoing are easily calculated mentally. Nevertheless it will be revealing to set out the gearing for all threads of from 6 to 15 tpi. A leadscrew of 8 tpi will be assumed:

$$6 \text{ T.P.I.} \quad \frac{\text{Driver}}{\text{Driven}} = \frac{8}{6} = \frac{40}{30} \quad \text{First driver} \atop \text{Leadscrew gear}$$

$$7 \quad \underline{\hspace{2cm}} = \frac{8}{7} = \frac{40}{35}$$

$$8 \quad \underline{\hspace{2cm}} = \frac{8}{8} = \frac{40}{40} \quad \text{(or any two} \atop \text{of equal size)*}$$

$$9 \quad \underline{\hspace{2cm}} = \frac{8}{9} = \frac{40}{45}$$

$$10 \quad \underline{\hspace{2cm}} = \frac{8}{10} = \frac{40}{50}$$

$$11 \quad \underline{\hspace{2cm}} = \frac{8}{11} = \frac{40}{55}$$

$$12 \quad \underline{\hspace{2cm}} = \frac{8}{12} = \frac{40}{60}$$

$$13 \quad \underline{\hspace{2cm}} = \frac{8}{13} = \frac{40}{65}$$

$$14 \quad \underline{\hspace{2cm}} = \frac{8}{14} = \frac{40}{70}$$

$$15 \quad \underline{\hspace{2cm}} = \frac{8}{15} = \frac{40}{75}$$

What should be noted in the list is that the driver remains at 40 throughout the range, and if this is replaced by a 20, then the threads/inch for which the lathe will be geared will be exactly double in each case. For example, the 9 tpi will increase to 18 tpi and the 13 to 26 tpi.

COMPOUND GEAR TRAINS

With a range of change wheels of from 20 to 75 teeth, the limit for simple reduction gearing consisting of one driver and one

---

* See also page 24 for equal ratio setting

22

driven, (or leadscrew gear) is reached at the 20:75 ratio, which, with a leadscrew of 8 tpi sets the lathe for cutting 30 tpi. Gearing for a greater number of threads/inch therefore calls for the use of "compound gearing".

One example of compound gearing is to be found in the wheels required for a thread of 40 to the inch. The same basic formula is used:

$$\frac{\text{Drivers}}{\text{Driven}} = \frac{\text{Threads/Inch of LS}}{\text{Threads/inch req.}}$$

and substituting the known figures for a leadscrew of 8 tpi we have:

$$\frac{\text{Drivers}}{\text{Driven}} = \frac{8}{40}$$

but if we now multiply 8 and 40 by 5, we get 40/200, and although this halves to 20/100, the 100 gear is outside our range. We therefore resolve 8/40 into factors:

$$\frac{\text{Drivers}}{\text{Driven}} = \frac{8}{40} = \frac{2 \times 4}{5 \times 8}$$

the factors are then raised to available change gear sizes by multiplying both 2 and 5 by 10; and 4 and 8 by 5:

$$\frac{\text{Drivers}}{\text{Driven}} = \frac{20}{50} \times \frac{20}{40}$$

The next question is, having found the gears, how are they set on the lathe?

What should be remembered here is that all gears in the numerator side are driving gears, and all gears in the denominator side are driven gears. It is generally necessary or more straightforward, however, to position the largest driven gear on the leadscrew, but provided the driven gears remain in a driven portion of the train the ratio will not be affected. Hence we may reverse or exchange the denominators to 20/40 × 20/50 and the gears would be set on the quadrant in the manner shown in the diagram, Fig. 8. Gear meshing limitations would prevent the direct engagement of the 20 first driver (No. 1) with the first driven 40 (No. 2), so the idle gear (A), here shown as a 65, is interposed. The 40 gear is coupled to the second 20 driver (No. 3) so that both revolve together, and the second 20 driver is then engaged with the 50 leadscrew wheel, (No. 4).

SCHEMATIC GEAR TRAIN PRESENTATION

The customary method for showing actual gear meshing sequences or arrangements

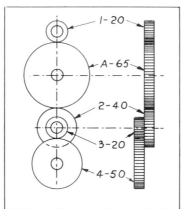
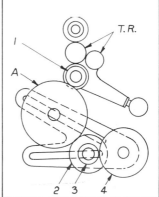

Fig. 8. An example of a single-compound gear train. Gearing shown is for cutting 40 tpi with an 8 tpi leadscrew.
At the left the gear train has been 'straightened out' for clarity. At the right, although the tumbler-reverse (TR) is shown, it plays no part in the final ratio.

in tabulated form calls for the use of fairly complicated headings to show not only the first driver and leadscrew gear, but whether or not the intermediate quadrant studs carry only an idler, or two wheels keyed together, as for compound trains. Thus the written layout of individual examples for explanatory purposes becomes sufficiently tedious as to discourage the presentation of more than an absolute minimum number, a circumstance which would interfere seriously with later discussions on gearing for metric pitches and allied matters.

With the foregoing disadvantages in mind a need was felt for a more straightforward method for indicating the actual positions of the gears on the lathe: a method which once explained would not call for the repetition of headings referring to first drivers, studs and leadscrew gear, or for any special mention of gears which are keyed together on the same quadrant stud.

The schematic method requires headings only for explanatory purposes, and here are three examples:

	First Driver	1st Quadrant Stud	2nd Quadrant Stud	Lead Screw
Simple	20— A — A —60			
Single Compound	20— A —40		20—50	
Double Compound	B — C	D — E	F — G	

In each case the lines connecting the gears show that the gears so joined are in direct mesh. Gears placed one above the other show that they are coupled or keyed together. Letter A is short for 'ANY', and

refers to any spare gear of suitable size that may be used as an 'idler' to connect main train gears that are too small to mesh directly together.

The gear at the extreme left is always the first driver, and the gear at the extreme right always the leadscrew wheel. Hence, if only one idle wheel had been used in the simple train, the layout would read:

$$20 — A — 60$$

It was convenient to use letters instead of numbers in terms of gear teeth to illustrate the double compound train because, when resolving a fractional solution into a practical layout it is necessary to ascertain that the sum of the number of teeth held by gears D and E exceeds the sum of C plus F by a minimum of five teeth otherwise C and F will mesh and will either lock the train solid, or prevent the proper meshing of D and E.

Sometimes it is useful to show the idle gears actually used, in which case the layout for 24 tpi with an 8 tpi leadscrew might read:

$$20 — 65 — 40 — 60$$

In a single compound train the idle gear may be placed between the first or second pair of main gears, and provided that the driving gears remain in a driving position, the ratio will not be altered:

$$20—50$$
$$20— A —40$$

will give exactly the same ratio as

$$20— A —40$$
$$20—50$$

Please also notice (1) a simple train such as

$$20 — A — A — 40$$

would be written for arithmetical checking purposes as

$$\frac{\text{Driver}}{\text{Driven}} = \frac{20}{40}$$

24

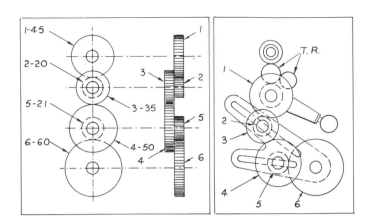

Fig. 8A. An example of a double-compound gear train. Gearing shown is for a metric pitch of 1.75 mm. to be cut from a leadscrew of 8 tpi. (This particular ratio calls for use of a 21T gear).

(2) The single compound train

$$20 - A - 40$$
$$20 - 50$$

$$\frac{\text{Drivers}}{\text{Driven}} = \frac{20}{40} \times \frac{20}{50}$$

(3) The double compound train

$$B - C \qquad F - G$$
$$\quad\; D - E$$

as $\dfrac{B}{C} \times \dfrac{D}{E} \times \dfrac{F}{G}$

B D F being drivers and *C E G* the driven gears. An example of a double compound train is given in Fig. 8A.

FURTHER CALCULATION NOTES

The calculation of change wheels for a thread of 19 to the inch is rather less obvious than the examples previously given. Assuming a leadscrew of 8 tpi we have:

$$\frac{\text{Drivers}}{\text{Driven}} = \frac{8}{19}$$

If we multiply 8 and 19 by 5 we get 40/95, which would serve well enough had we a 95 change gear, but we assume that the set stops at 75 teeth. What should be noted now is that if no wheel is available that is an exact multiple of 19, then precision gearing is impossible. However, the change gear set will probably include a special wheel of 38 teeth, whereupon our initial formula will read:

$$\frac{\text{Drivers}}{\text{Driven}} = \frac{8}{38} \times \frac{2}{1}$$

where the multiplication by 2/1 is simply written in to hold the ratio. Proceeding from here, if the 8 and 1 are now multiplied by 5 we have

$$\frac{\text{Drivers}}{\text{Driven}} = \frac{40}{38} \times \frac{2}{5}$$

and finally, multiplying 2 and 5 by 10 gives:

$$\frac{\text{Drivers}}{\text{Driven}} = \frac{40}{38} \times \frac{20}{50}$$

which could be set on the lathe:

$$20 - A - 38$$
$$40 - \quad 50$$

In the example just given, 19 is, of course, a prime number, and the impossibility of *exactly* gearing a lathe for primes or multiples of primes should be noted. Thus with change gears by fives, 11 and 22 tpi call for a 55 change gear,

25

13 and 26 tpi call for a 65, although it can be an advantage to have special gears of 33 or 44, and 39 teeth. We may note, too, that for example a pitch of 1.1 mm requires a 55 wheel, and 0.65 mm would call for a 65 gear, in the event of its being necessary to cut these non-standard pitches with (theoretical) accuracy.

MIXED NUMBERS

Occasionally it is necessary to gear a lathe for cutting a mixed number of threads to the inch. As an example of this, let us assume that gearing for a thread of $10\frac{1}{2}$ tpi is to be cut from a leadscrew of 8 tpi:

$$\frac{\text{Drivers}}{\text{Driven}} = \frac{8}{10\frac{1}{2}}$$

Multiply both numerator and denominator by 2 to eliminate the awkward fraction = 16/21 which factorises to 4/3 × 4/7. Multiply the first 4 and 3 by 10, and the second 4 and 7 by 5 and we have drivers/driven = 40/30 × 20/35 which would set on the quadrant, e.g.:

$$20 — A — 30$$
$$40 — 35$$

However, as will be explained, larger gearing offers a more mechanically sound gear train, and it would be as well to increase the 20/30 ratio to 30/45 (by multiplying 2/3 by 15) thus offering a quadrant setting:

$$30 — A — 45$$
$$40 — 35$$
$$\text{or} \quad 40 — A — 35$$
$$30 — 45$$

Note that when we say, e.g. 'multiply 2/3 by 15' to bring to change gear sizes, we are using shorthand for 'multiply both numerator and denominator by 15', and are thus taking a mathematical liberty because, of course, 2/3 × 15 = 10.

Some mixed number threads/inch resolve into primes, e.g. $11\frac{1}{2}$ tpi = prime

23 whole threads in 2 in., hence accurate gearing cannot be set unless we have a 23 or 46 gear. The gearing required is Drivers/Driven = $8/11\frac{1}{2}$ = 16/23 = 4/23 ×4/1. Multiply the first 4 and last 1 by 5 = 20/23 × 4/5 = 20/23 × 40/50, or, to avoid use of small gears, double 20/23 to 40/46, and if two 40's are not available, multiply 4/5 by 15, giving 60/75, thus offering a complete gear train: 40/46 × 60/75 which would set on the quadrant e.g. –

$$40 — A — 46$$
$$60 — 75$$

However, if a 23 or 46 gear is not available, and if a small pitch error is not objected to, then suitable gearing using available gears by fives can be found by approximation, as explained in Section 4 which shows that a quadrant setting:

$$30 — A — 40$$
$$65 — 70$$

would serve.

EQUAL RATIO SETTING

As has already been explained, when it is necessary to gear a lathe for cutting a thread of the same pitch as that of the leadscrew, the leadscrew must revolve at the same speed as that of the component to be threaded.

If it is agreed that there are two 20 teeth change gears included with the set, then these, of course, may be arranged:

$$20 — A — A — 20$$

but in consideration of the fact that the smaller the gear diameter, the greater the loading on the teeth (compare the action of short and long levers) the use of gears, which if of No. 20 diametral pitch have a radius of action of only half an inch, would seem to contain an element of mechanical unsoundness, especially as the pitch to be cut will be fairly coarse and inadvertent

26

leadscrew loading could be quite high. Agreed the tumbler-reverse of some small lathes transmits power through pinions of small diameter. but one may as well avoid a repetition of the arrangement if possible. The most straightforward approach therefore is to obtain an extra gear having, say, 40 teeth, and to set the lathe:

$$40 - A - A - 40$$

which doubles the teeth velocity and halves the teeth loading without altering the torque on the driving and driven keyways.

Another way of obtaining an equal ratio when two gears of equal size are not available is to use a single compound gear train containing first a ratio increase, and second, the same ratio inverted, such as

$$\frac{45}{30} \times \frac{40}{60}$$

which reduces to

$$\frac{3}{2} \times \frac{2}{3} = \frac{1}{1}$$

and which would set on the lathe:

$$40 - A - 30$$
$$45 - 60$$

For the convenience of those who may like a quick reference, Table T1 gives gearing for threads/inch from a leadscrew of 8 tpi, and Table T2 gives gearing for threads/inch from a leadscrew of 10 tpi.

THREADS DESIGNATED BY LEAD

Occasionally a thread is designated by lead instead of by threads/inch or by pitch. (Lead is the distance a screw advances axially in one turn in a fixed nut). It is necessary to designate multiple-start threads by lead to distinguish the axial advance of the screw from the pitch, which, with any given lead, will decrease in proportion to the number of starts (see section 6)

With English measure, lead is sometimes expressed in fractional form, and may be given as, for example, 5/32 inch.

One way of handling the fraction for gearing calculations is to first convert to threads per inch:

$$\frac{1}{5/32} = \frac{1}{1} \times \frac{32}{5} = \frac{32}{5} = 6\frac{2}{5} \text{ T.P.I.}$$

or 32 whole threads in 5 inches.

With a leadscrew of 8 tpi a length of 5 inches contains 40 leadscrew threads so we have:

$$\frac{\text{Drivers}}{\text{Driven}} = \frac{8}{32} \times \frac{5}{1} = \frac{40}{32}$$

and, assuming a 32 is not available, divide 40 and 32 by 8:

$$= \frac{5}{4} = \frac{50 \text{ Driver}}{40 \text{ Driven}}$$

Another approach lies in the expression of both leadscrew and lead of the work in terms of lead, giving a formula:

$$\frac{\text{Drivers}}{\text{Driven}} = \frac{\text{Lead of screw to be cut}}{\text{Lead of leadscrew}}$$

$$\frac{\frac{5}{32}}{\frac{1}{8}} = \frac{5}{32} \times \frac{8}{1} = \frac{5 \text{ Driver}}{4 \text{ Driven}}$$

Evidently, now, either of the foregoing gives the same result as the more direct:

$$\frac{\text{Drivers}}{\text{Driven}} = \frac{\text{Numerator}}{\text{Denominator}} \times \frac{\text{T.P.I. of LS}}{1}$$

Screws having a lead greater than that of the leadscrew require that the leadscrew shall revolve at a higher rate than the work, and although a small lathe with an 8 tpi leadscrew would handle a lead of 5/32 inch well enough, the strain upon the 3:1 step-up gearing required for a lead of $\frac{3}{8}$ inch, for example, would be severe indeed, especially as the efficiency of a leadscrew and nut as a transmitter of power can be as low as 30 to 40 percent. To reduce the strain, some early text

Table T1. Gearing for threads/inch with a leadscrew of 8 tpi. Assumes gears 20-20-75 by fives are available. Alternative gearing for 11½ and 19 tpi requires a 38T gear.

THREADS PER INCH REQ.	QUADRANT GEARING	THREADS PER INCH REQ.	QUADRANT GEARING
4	40 —— A —— A —— 20	19_B	20 —— A —— $\frac{38}{40}$ —— 50
5	40 —— A —— A —— 25	20	20 —— A —— A —— 50
6	40 —— A —— A —— 30	22	20 —— A —— A —— 55
7	40 —— A —— A —— 35	24	20 —— A —— A —— 60
8_A	40 —— A —— A —— 40	25	40 —— A —— $\frac{50}{30}$ —— 75
8_B	40 —— A —— $\frac{30}{45}$ —— 60	26	20 —— A —— A —— 65
9	40 —— A —— A —— 45	28	20 —— A —— $\frac{35}{20}$ —— 40
10	40 —— A —— A —— 50	32	20 —— A —— $\frac{40}{25}$ —— 50
11	40 —— A —— A —— 55	36	20 —— A —— $\frac{40}{20}$ —— 45
$11\frac{1}{2}A$	40 —— A —— $\frac{50}{65}$ —— 75	40	20 —— A —— $\frac{40}{20}$ —— 50
$11\frac{1}{2}B$	$\frac{55 - 25}{20 - 50}$ $\frac{30 - 38}{50}$	44	20 —— A —— $\frac{50}{25}$ —— 55
12	40 —— A —— A —— 60	48	20 —— A —— $\frac{50}{25}$ —— 60
13	40 —— A —— A —— 65	56	20 —— A —— $\frac{50}{25}$ —— 70
14	40 —— A —— A —— 70	60	20 —— A —— $\frac{50}{20}$ —— 60
16	20 —— A —— A —— 40	64	$20 - \frac{40}{30}$ $\frac{25}{50} - 60$
18	20 —— A —— A —— 45	72	$20 - \frac{45}{30}$ $\frac{20}{40} - 60$
19_A	35 —— A —— $\frac{50}{45}$ —— 75	80	$20 - \frac{40}{30}$ $\frac{25}{50} - 75$

28

Table T2. Gearing for threads/inch with a leadscrew of 10 tpi. Assumes gears 20-20-75 by fives are available. Both sets of gearing for 11½ tpi are approximations. B is the best, but requires a 38T gear. Of the two sets of gearing for 19 tpi, A is an approximation, B is correct.
For a leadscrew of 5 tpi (1) halve any driver or (2) double any driven gear, or (3) include a 1 to 2 reduction ratio.

THREADS PER INCH REQ	QUADRANT GEARING		THREADS PER INCH REQ	QUADRANT GEARING
4	50 ── A ── A ── 20		19B	20 ── A ── A ── 38
5	50 ── A ── A ── 25		20	20 ── A ── A ── 40
6	50 ── A ── A ── 30		22	25 ── A ── A ── 55
7	50 ── A ── A ── 35		24	20 ── A ── 30 / 25 ── 40
8	50 ── A ── A ── 40		25	20 ── A ── A ── 50
9	50 ── A ── A ── 45		26	25 ── A ── A ── 65
10A	40 ── A ── A ── 40		28	20 ── A ── 40 / 25 ── 35
10B	40 ── A ── 30 / 45 ── 60		32	25 ── A ── 40 / 30 ── 60
11	50 ── A ── A ── 55		36	20 ── A ── 40 / 25 ── 45
11½A	40 ── A ── 50 / 65 ── 60		40	20 ── A ── 50 / 25 ── 40
11½B	30 ── A ── 38 / 55 ── 50		44	20 ── A ── 40 / 25 ── 55
12	25 ── A ── A ── 30		48	20 ── A ── 40 / 25 ── 60
13	50 ── A ── A ── 65		56	20 ── A ── 40 / 25 ── 70
14	25 ── A ── A ── 35		60	20 ── A ── 50 / 25 ── 60
16	25 ── A ── A ── 40		64	30 ── 40 / 25 ── 40 20 ── 60
18	25 ── A ── A ── 45		72	25 ── A ── 60 / 25 ── 75
19A	35 ── A ── 40 / 30 ── 50		80	25 ── 40 / 20 ── 60 30 ── 50

books recommend that the leadscrew itself be driven, thus reversing the function of the gearing from a step-up to a step-down ratio driving the component to be threaded. No doubt in the early days when the mechanisms of all lathes were exposed, such an arrangement was not too difficult. For a small lathe such as the Myford $3\frac{1}{2}$ in. some means would have to be devised for by-passing the tumbler-reverse, otherwise the small pinions thereon would have to transmit sufficient torque to rotate the lathe spindle against the drag of a threading tool.

With modern lathes the objection to high ratio step-up gearing for coarse leads can be overcome by two methods.

In one, special headstock gearing allows of the work being driven at reduced speed while the leadscrew gearing is driven from a higher speed element in the headstock. For example, if one of the 'back-gear' speeds is six times as slow as the ungeared drive and the change gear train is driven from the ungeared element, a leadscrew driven through 1:1 gearing would be revolving at six times the work speed. If the leadscrew was of 4 tpi then six revolutions would advance the carriage through $1\frac{1}{2}$ inches for each revolution of the work. In the unlikely event of a lathe with this feature not being provided with a fully selective gearbox, calculations for other leads with the special drive in use would be made on the assumption that the leadscrew was of $1\frac{1}{2}$ inch lead, or 2/3 tpi. Thus for example, gearing for a lead of 3/4 in. would be calculated as follows:

$$\frac{\text{Drivers}}{\text{Driven}} = \frac{\text{Lead of screw to be cut}}{\text{Lead of leadscrew}}$$

Hence —

$$\frac{\text{Driver}}{\text{Driven}} = \frac{\frac{3}{4}}{\frac{3}{2}} = \frac{3}{4} \times \frac{2}{3} = \frac{1}{2} \text{ or } \frac{20}{40}$$

Similar results are sometimes obtained by use of a speed-reducing chuck.

CHECKING

The correctness of any given change gear set-up or calculated ratio for cutting a particular number of threads to the inch may be checked by inverting the gear train in its fractional form and multiplying by the number of threads per inch of the leadscrew:

$$\text{TPI} = \frac{\text{Driven}}{\text{Drivers}} \times \frac{\text{LS TPI}}{1}$$

For example, the gearing needed to cut 28 tpi with an 8 tpi leadscrew may have been calculated to:

$$\frac{\text{Drivers}}{\text{Driven}} = \frac{20}{35} \times \frac{20}{40}$$

and a check is desired. Inserting the figures in the above formula we have:

$$\text{TPI} = \frac{35}{20} \times \frac{40}{20} \times \frac{8}{1} = 28$$

If, instead of threads/inch, it is desired to find the pitch of a thread which will be cut by a given gear combination, then the gear train is left in its original fractional form and multiplied by the *pitch* of the leadscrew:

$$\text{Pitch} = \frac{\text{Drivers}}{\text{Driven}} \times \frac{1}{\text{LS TPI}}$$

Thus the pitch produced by the foregoing gear train would be:

$$\frac{20}{35} \times \frac{20}{40} \times \frac{1}{8} = \frac{1}{28} = 0.0357142$$

With an 8 tpi leadscrew, what pitch will be given by the following gear train which offers an approximation to a thread of $11\frac{1}{2}$ to the inch?

$$40 - A - 50$$
$$65 - 75$$

Here the gear train must first be rewritten in fractional form, then multiplied by leadscrew pitch:

$$\frac{\text{(Drivers)}}{\text{(Driven)}} = \frac{40}{50} \times \frac{65}{75} \times \frac{1}{8} = \frac{13}{150}$$

$$= 0.086666 \text{ inch.}$$

The actual pitch of $11\frac{1}{2}$ tpi is 0.0869565 inch so the above gear train would produce a minus pitch error of 0.0869565 minus 0.086666 = 0.00029 inch, say 3 tenths of a thousandth of an inch, or 0.00762 mm.

SELF-ACT FEEDS

When a small lathe is not fitted with a separate feed shaft and the leadscrew has to be used for self-act, or slow carriage traversing purposes, and is being driven through a double compound reduction gear train, it is sometimes useful to know by how much the carriage advances for each revolution of the work being turned. This is the same as asking for what pitch the lathe is geared, and the pitch formula is therefore used:

$$\text{Pitch} = \frac{\text{Drivers}}{\text{Driven}} \times \frac{1}{\text{LS TPI}}$$

As an example, suppose the following gearing was being used to drive a lead-screw of 8 tpi:

$$20 - 65 \quad 20 - 75$$
$$25 - 70$$

then pitch $=$

$$\frac{\text{Drivers}}{\text{Driven}} = \frac{20}{65} \times \frac{25}{70} \times \frac{20}{75} \times \frac{1}{8}$$
$$= \frac{1}{273} = 0.003663 \text{ inch per revolution}$$

of the work, or 273 tpi or "cuts per inch"

DIAMETRAL PITCH WORMS

It is sometimes necessary to use a lathe to cut a worm. From the experimenter's point of view, a worm may be required to mesh with one of the lathe change gears, although for the worm to mesh at right angles to the axis of the worm wheel, if this is an ordinary spur gear, the body of the worm thread has to be undersized, or "thinned".

For purposes of calculating the lathe quadrant gearing for a worm it is not necessary to find the actual number of threads to the inch which would give a satisfactory meshing: all that need be known is the number of threads to the inch of the leadscrew and the diametral pitch (DP) of the gear with which the worm is to mesh.

The diametral pitch number of a gear is that number which states how many teeth the gear holds to each inch of (pitch) diameter, and the DP of any gear presumed to be of English origin (i.e. sized by inch measure) may be found in the following way:

1. Measure the whole, or outside diameter.
2. Count the number of teeth and add 2.
3. Divide the teeth total so found by the whole diameter.

For example a gear has an outside diameter of 3.100 inches, and has 60 teeth. Add 2, giving 62, and divide by 3.100. The figure 20 so obtained will be the DP of the gear.

To give one more example: a gear has an outside diameter of 2.625 inches, and has 40 teeth. The DP therefore equals 42 divided by 2.625 = 16.

The foregoing are of course perfect examples wherein the gear diameter is assumed to be correct, whereas in practice an undersizing would be most probable, in which case a fractional or mixed number DP would result from the calculation. However, the undersizing of the diameter would be unlikely to be so severe as to cause doubt as to whether a gear was, for example, of No. 16 or 17 DP, and the nearest whole number may therefore be taken.

The drawing Fig. 9 illustrates the chief dimensional terms used in sizing a spur

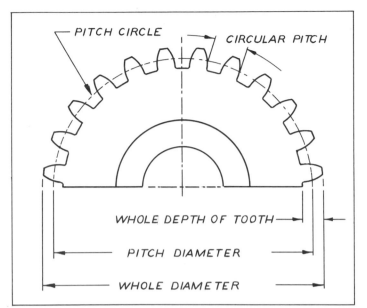

Fig. 9. Showing the terms used in sizing a gear.

gear. Although a spur gear differs greatly in tooth form from the teeth of a true worm wheel, a description of which is outside the scope of this work, the spur gear will serve well enough for explaining the terms associated with a worm thread, which calls for an understanding of the diametral and circular pitch.

The pitch diameter, about which the teeth are formed, is that diameter which would decide the ratio between two plain cylinders in frictional contact.

The circular pitch is the distance between any one point on one tooth and the corresponding point on an adjacent tooth, measured along the pitch circle, and may be found from 3.1415926/ diametral pitch.

The whole depth of a tooth space, or depth of a worm thread, may be found from:

$$D = \frac{2.157}{DP}$$

FORMULA FOR DP WORM THREADS

Taking Pi as 3.1415926, the formula for gearing a leadscrew of ANY threads/inch value for cutting worm threads sized by diametral pitch would read:

$$\frac{\text{Drivers}}{\text{Driven}} =$$

$$\frac{\text{Leadscrew threads/inch}}{\text{Diametral pitch of worm}} \times \frac{3.1415926}{1}$$

This of course, although supremely accurate, is rendered impracticable by the 7-digit decimal portion for Pi, consequently Pi has to be converted, by approximation, into an improper fraction wherein the numerator and denominator figures will lend themselves to conversion into available change gears. A list of 15 such approximations for Pi will be found in the accompanying Table T3. Here the reader will notice that of the 15 approximations,

32

TABLE No. T3
LIST OF APPROXIMATIONS FOR Pi
(Based on Pi = 3.1415926)

	Approximation	Error* +/− 1 part in parts:		Approximation	Error* +/− 1 part in parts:
1.	$\dfrac{5 \times 71}{113}$	_____	9.	$\dfrac{7 \times 35}{6 \times 13}$	− 5527
2.	$\dfrac{47 \times 127}{38 \times 50}$	−212 765	10.	$\dfrac{10 \times 38}{11 \times 11}$	−2261
3.	$\dfrac{13 \times 29}{8 \times 15}$	+43 103	11.	$\dfrac{22}{7}$	+2486
4.	$\dfrac{9 \times 37}{2 \times 53}$	−37 313	12.	$\dfrac{5 \times 27}{1 \times 43}$	−1525
5.	$\dfrac{23 \times 28}{5 \times 41}$	−24 096	13.	$\dfrac{5 \times 59}{2 \times 47}$	− 953
6.	$\dfrac{19 \times 21}{127}$	+22 675	14.	$\dfrac{5 \times 32}{3 \times 17}$	− 724
7.	$\dfrac{17 \times 17}{4 \times 23}$	−10 857	15.	$\dfrac{3 \times 23}{2 \times 11}$	− 600
8.	$\dfrac{15 \times 31}{4 \times 37}$	+10 537	16.	$\dfrac{25}{8}$	− 189

*The error figures are as obtained from an 8-digit calculator which cannot reveal precise errors when these are minimal.

Table 3A. Change gears for cutting threads sized by diametral pitch with a leadscrew of 8 tpi and change gears 'by fives'. Train at right for No. 14 DP requires two 35T gears.

D.P. NUMBER	BASED ON PI = $7/6 \times 35/13$			BASED ON PI = $22/7$	
	TRUE CIRCULAR PITCH INCH MEASURE	QUADRANT GEARING	PITCH ERROR MILLIONTHS INCH	QUADRANT GEARING	PITCH ERROR MILLIONTHS INCH
14	0.2243994	50 —— A —— 30/70 —— 65	−40.5	40 —— A —— 35/55 —— 35	+90.3
16	0.1963495	50 —— 40/70 —— A —— 35/30 —— 65	−35.5	55 —— A —— 35/55 —— 35	+79.0
18	0.1745329	50 —— 45/70 —— A —— 35/30 —— 65	−31.5	40 —— A —— 35/55 —— 45	+70.2
20	0.1570796	35 —— A —— 30/70 —— 65	−28.4	40 —— A —— 35/55 —— 50	+63.2
22	0.1427996	50 —— 55/70 —— A —— 35/30 —— 65	−25.8	40 —— A —— 35/55 —— 35	+57.5
24	0.1303996	50 —— 45/70 —— A —— 35/40 —— 65	−23.6	40 —— A —— 35/55 —— 60	+52.7
26	0.1208304	35 —— A —— 39/70 —— 65	−21.8	40 —— A —— 35/55 —— 65	+48.7
28	0.1121997	50 —— A —— 60/70 —— 65	−20.3	40 —— A —— 35/55 —— 70	+45.1
30	0.1047197	35 —— A —— 45/70 —— 65	−18.9	40 —— A —— 35/55 —— 75	+42.2
32	0.0981747	50 —— 60/70 —— A —— 35/40 —— 65	−17.7	20 —— A —— 40/55 —— 35	+39.5
36	0.0872664	50 —— 45/70 —— A —— 35/60 —— 65	−15.7	20 —— A —— 45/55 —— 35	+35.1
40	0.0785398	35 —— A —— 60/70 —— 65	−14.2	20 —— A —— 35/55 —— 50	+31.6

only two lend themselves to use with gears 'by fives' and show errors that are not altogether unreasonable: these are No. 9 with an error of minus one part in 5527 parts, and No. 11, the well-known 22/7 with an error of plus one part in 2486 parts.

However, with the list of Pi approximations on hand, our formula for DP worms will now read:

$$\frac{\text{Drivers}}{\text{Driven}} = \frac{\text{Leadscrew tpi}}{\text{DP of worm}} \times \text{Pi approx.}$$

Hence, if a leadscrew is of 8 tpi and we wish to cut a worm to mesh with Myford change gears which are of No. 20 DP and we select No. 11 approximation for Pi, substituting the known figures we have:

$$\frac{\text{Drivers}}{\text{Driven}} = \frac{8}{20} \times \frac{22}{7}$$

wherein 22 and 20 are divisible by 2, giving 8/10 × 11/7, and multiplying all figures by 5, we have 40/50 × 55/35, which would set on a quadrant:

$$40 - A \quad - \quad 35$$
$$55 - 50$$

Similarly it can be shown that if we use No. 9 approximation for Pi, No. 20 DP from a leadscrew of 8 tpi will require gearing:

$$35 - A \quad - \quad 30$$
$$70 - 65$$

It may occasionally be useful to note that with a leadscrew of 8 tpi, gearing 50/40 will cut a No.20 DP worm with a theoretical pitch error of minus 0.00083 in. (This gearing shortens 'pick-up' – see Section 5) However, the 50/40 ratio uses Pi approximation No. 16 (from the Table T3) which itself has a very high error.

To give one more example of a gearing calculation, let us assume a worm is required to mesh with a gear of No. 16 DP, and that a leadscrew has 10 threads/inch. If we select No. 11 approximation for Pi, we have:

$$\frac{\text{Drivers}}{\text{Driven}} = \frac{10}{16} \times \frac{22}{7} = \frac{5}{4} \times \frac{11}{7}$$
$$= \frac{50}{40} \times \frac{55}{35}$$

which would set on a quadrant e.g.:

$$50 - A \quad - \quad 40$$
$$55 - 35$$

It should be noted, however, that the lower the DP number, the more coarse the thread. In this example there is an approximate 2 to 1 step-up in the leadscrew speed relative to that of the component being threaded, and the strain on a small lathe and its gearing would be significant.

The Table T3A offers two sets of gearing for cutting worms sized by diametral pitch from a leadscrew of 8 threads/inch and gears 'by fives'. The gearing at the left is derived from No. 9 approximation, and the right-hand gearing from No. 11 approximation for Pi. For a leadscrew of 4 threads/inch, halve any driving gear, or double any driven gear.

It is worth noting that No. 3 approximation for Pi (Table T3) is usable if a 29T special gear is available. For example, gearing for No. 20 DP from a leadscrew of 8 tpi resolves to a neat:

$$29 - A \quad - \quad 30$$
$$65 - 50$$

which under perfect conditions would offer a worm pitch error of only plus 3.7 millionths/inch.

Of the Pi approximations in general, although Nos. 2 and 6, which each contain the high prime 127, may appear practically useless, the 127's cancel when used in conjunction with gearing for diametral pitch from a metric leadscrew (No. 6 approx) or for gearing for module (qv) from a threads/inch leadscrew for which No. 2 Pi approximation can be used.

CHECKING PROPOSED QUADRANT GEARING

To check any proposed gearing it is advisable to use the ordinary pitch-checking formula: Drivers/Driven × leadscrew pitch, and to note how the resulting figures compare with the proper circular pitch of the worm as found from the formula:

$$\text{Circular pitch} = \frac{3.1415926}{\text{Diametral Pitch}}$$

Thus, the pitch given by gearing 40/35 × 55/50 for No. 20 DP cut from a leadscrew of 8 tpi is:

$$\frac{40}{35} \times \frac{55}{50} \times \frac{1}{8}$$

$$= 0.1571428 \text{ inch}$$

and as No. 20 DP CP = 0.1570796 in. there is an error of plus 63.2 millionths/inch.

Similarly, the circular pitch given by gearing an 8 tpi leadscrew:

$$29 - A - 30$$
$$65 - 50$$

will be 29/30 × 65/50 × 1/8 = 0.1570833 inch, showing a plus error of 0.0000037 in. on true CP of 0.1570796 in.

If a knowledge of the threads/inch of any worm is required, this can be found from:

$$(1) \text{ Threads/inch} = \frac{1}{\text{Circular Pitch}}$$

or (2) from the gearing

$$\text{Threads/inch} = \frac{\text{Driven}}{\text{Drivers}} \times \frac{\text{Leadscrew tpi}}{1}$$

Method (2) is the more revealing when resolved to a mixed number fraction. For example the threads/inch given by gearing 40/35 × 55/50 with a leadscrew of 8 tpi = 35/40 × 50/55 × 8 = 70/11 = 6 & 4/11.

Method (1) of course gives the 'true' threads/inch value to close limits, and method (2) gives the threads/inch that would be produced by the gearing under perfect conditions.

LEADSCREW ACCURACY

Of course, unless special provision is made, the pitch accuracy of any thread or worm thread cut in a lathe cannot exceed the accuracy of the leadscrew used, yet in this respect there seems to be a strongly held notion that lathe leadscrews are of correct pitch to very close limits, whereas, except by accident, this is most unlikely. Indeed even if a leadscrew was initially of high accuracy it would be unlikely to remain in that condition for any worthwhile period because of its exposed position. No doubt lathe makers would assert that their leadscrews are reasonably accurate, but what is reasonable? A pitch error of minus 2/10 thou. inch on a leadscrew of 4 tpi becomes minus 8/10 thou. in. over a 1 in. length, and 2.4 thou. inches over a 3 in. length.

One of the author's 8 tpi leadscrews has an error averaging minus 0.0005 in. per inch, or minus 0.006 in. per foot, and in gearing for some 40 tpi micrometer-type threads for his own use, which required a pitch error not exceeding 0.0001 in. per inch, it was necessary to modify the gearing to speed-up the 'slow' leadscrew in the ratio 2001 to 2000. This ratio is offered by:

(A) 29/25 × 23/20 × 3/4

and when this is included in the basic 1 to 5 reduction for 40 tpi with a leadscrew of 8 tpi, the following quadrant gearing is given:

$$45 - 40 \qquad 23 - 75$$
$$29 - 50$$

This gearing with a perfect leadscrew would cut a pitch of 0.0250125 in., but of course, as the leadscrew was 'slow' the last three decimal places were eliminated, leaving the required 0.025 in. pitch.

It might also be handy to note that a ratio of 2002 to 2001 is given by:

(B) $14/29 \times 13/23 \times 11/3$

And, of course, either (A) or (B) inverted will give a corresponding small speed reduction.

The Table T3B (compiled by the author) gives a selection of 48 close ratios from 125/126 to 6003/6004, factorised into 2 and 3-element ratios for use when necessary for thread pitch modification or experimental purposes.

SELECTIVE GEARBOXES

Incidentally, it is the inability to manipulate change gears on lathes with built-in selective 'threading' gearboxes that led to the author objecting to such arrangements. With selective gearboxes, of course, the makers decide what pitches shall be available, and immediately an unusual pitch is required it cannot be set without the protracted procedure of writing to the manufacturers with a request for the necessary gears, which, upon receipt, are set on a quadrant – often termed by the makers 'outside gearing'. Commercially, in a majority of instances, the necessity for special ordering leads to a refusal to process one or a few special threads: a curious reflection on the lathe which was once known as 'The King of Tools'.

Table T3B. A selection of 48 close ratios from 125/126 to 6003/6004 factorised into 2 & 3 – element ratios for use when necessary for thread pitch modification or experimental purposes.

$125/126 = 5/6 \times 25/21$	$2744/2745 = 49/45 \times 56/61$
$252/253 = 12/11 \times 21/23$	$3002/3003 = 38/39 \times 79/77$
$505/506 = 5/11 \times 101/46$	$3248/3249 = 56/57 \times 58/57$
$702/703 = 18/19 \times 39/37$	$3255/3256 = 35/44 \times 93/74$
$703/704 = 19/22 \times 37/32$	$3509/3510 = 11/9 \times 11/10 \times 29/39$
$704/705 = 22/15 \times 32/47$	$3519/3520 = 51/55 \times 69/64$
$749/750 = 7/25 \times 107/30$	$3750/3751 = 5/11 \times 15/11 \times 50/31$
$759/760 = 23/20 \times 33/38$	$3751/3752 = 11/7 \times 11/8 \times 31/67$
$799/800 = 17/20 \times 47/40$	$3999/4000 = 43/50 \times 93/80$
$800/801 = 20/9 \times 40/89$	$4004/4005 = 7/5 \times 22/9 \times 26/89$
$850/851 = 25/23 \times 34/37$	$4255/4256 = 5/7 \times 23/16 \times 37/38$
$999/1000 = 27/20 \times 37/50$	$4484/4485 = 59/65 \times 76/69$
$1000/1001 = 20/13 \times 50/77$	$4488/4489 = 68/67 \times 66/67$
$1247/1248 = 29/39 \times 43/32$	$4514/4515 = 2/5 \times 37/21 \times 61/43$
$1500/1501 = 30/19 \times 50/79$	$4752/4753 = 48/49 \times 99/97$
$1504/1505 = 32/35 \times 47/43$	$4756/4757 = 58/67 \times 82/71$
$1749/1750 = 33/35 \times 53/50$	$4991/4992 = 7/8 \times 23/16 \times 31/39$
$1750/1751 = 35/17 \times 50/103$	$5015/5016 = 59/57 \times 85/88$
$2000/2001 = 5/3 \times 8/23 \times 50/29$	$5247/5248 = 53/64 \times 99/82$
$2001/2002 = 3/11 \times 23/13 \times 29/14$	$5250/5251 = 5/1 \times 21/59 \times 50/89$
$2254/2255 = 46/41 \times 49/55$	$5499/5500 = 3/5 \times 39/20 \times 47/55$
$2255/2256 = 41/47 \times 55/48$	$5529/5530 = 57/70 \times 97/79$
$2499/2500 = 49/50 \times 51/50$	$5750/5751 = 5/1 \times 23/71 \times 50/81$
$2500/2501 = 50/41 \times 50/61$	$6003/6004 = 9/4 \times 23/19 \times 29/79$

Admitted, quick change gearboxes do allow for the rapid selection of any pitch within the range, yet, at the time of writing, with a majority of centre lathes, the time saved is largely negated by the archaic means provided for holding pick-up, (avoidance of 'crossed thread' during screwcutting) and the slowness at which the required pitch has to be cut.

750 THOUSAND PITCHES

According to Prof. D. H. Chaddock, twenty change gears can be set in 380 2-gear combinations, 29,070 4-gear combinations, and 775,200 6-gear combinations, a total of 804,650 ways, and allowing 10,000 or so for identical ratios arising from the assembly of the same gears in a different order, the total possible *ratios* from 20 gears in 2, 4 and 6-gear combinations is about three quarters of a million, but not if they are locked up in a gearbox. Instead therefore of increasing the versatility of a lathe, a selective screwcutting gearbox reduces it.

EVOLUTION OF GEARBOX

The evolution of the quick selective threading gearbox can be traced back to the early screwcutting lathes which were provided only with a leadscrew, quadrant, and set of change gears, and these gears were set on the quadrant either for screwcutting ratios – generally calling for a simple train of gears – or for self-act feed rates, which required a compound or double compound train of gears to give a sufficiently slow leadscrew rotation. Consequently when a quadrant was set for screwcutting, no self-act rate was available, and vice-versa if a quadrant was set for self-act. Thus it was impossible to alternate quickly between screwcutting and self-act feed rates, and it was such a time-consuming operation to change the quadrant gearing between self-act and screwcutting that users demanded the lever operated selected gearbox, lumping together a selection of screwcutting and feed ratios, but failing completely to perceive that the only genuine requirement was for some sort of slow drive for the feedshaft or leadscrew, this drive being totally *independent* of any gearing that may have been set on a quadrant for screwcutting purposes, and always available even when a change gear quadrant was vacant. And of course, such an independent self-act does offer immediate change to or from screwcutting ratios. The author first fitted independent self-act reduction gearing in 1940, and when a change gear quadrant is uncluttered with double-compound reduction gearing for self-act, a majority of common thread pitches can be set in less than 60 seconds by merely substituting one gear for another and adjusting the new meshing distances. Hardinge use an infinitely variable speed electric motor for self-act on their HLV-H High Precision Lathes, and this has the additional advantage that feed rates can be varied whilst a cut is in progress. Hardinge do however provide a selective gearbox for most-used threads. Indeed prospective industrial lathe buyers would today look somewhat askance at any lathe without a 'threading box', so ingrained is the belief that this refinement is of paramount importance.

SELECTIVE GEARBOXES FOR MODEL MAKERS' LATHES

Those contemplating purchase of a small lathe may like to note that if a selective threading gearbox is felt desirable, then:
(1) A decision should be made whether to order a metric or an English leadscrew and gearbox. (If, of course, there is a choice).

(2) Whichever is chosen, extra gears called a 'conversion set' would have to be purchased for cutting threads of opposite language to that of the leadscrew.

(3) When a 'conversion set' is in use for cutting threads of opposite language to that of the leadscrew, self-act, or automatic slow carriage traverse will not be available until after removal of the conversion gearing and replacement of the 'normal' gears used for connecting the lathe spindle to the gearbox input.

(4) Thread pitches of a coarseness much exceeding that of the leadscrew cannot be cut without danger of overstraining the gearbox gears.

(5) A good range of spare change gears for cutting odd thread pitches outside the range offered by the gearbox, (or even outside the range offered by the gearbox together with the conversion set) and for indexing and general experimental work, will not be available unless purchased separately.

SECTION 3

Gearing an English leadscrew for metric threads

Although there were slight differences between the English and American ratios for the number of millimetres to the inch, the difference was so small as to be of interest only to the manufacturers of precision instruments and gauge blocks, the latter generally being sized to a millionth of an inch, or to 25 millionths of a millimetre. Nowadays, however, it has been generally agreed that one inch contains exactly 25.4 mm, or 2.54 centimetres.

Perhaps the simplest way to show, basically, how a lathe with an English leadscrew may be geared to cut a metric thread is by the following reasoning. Suppose, for example, a lathe with a leadscrew of 8 tpi is geared to cut a thread of 20 tpi with a basic gearing ratio of 8/20, and we wish to modify the gearing to cut 20 threads per centimetre (0.5 mm pitch) instead of 20 tpi. Now, 20 threads per CM is, of course, 'finer' than 20 tpi, and in fact, as there are 2.54 CM to the inch, 20 threads/CM is 20 × 2.54 = 50.8 threads to the inch, or 0.5 mm pitch. Accordingly, to gear for the finer pitch, the 8/20 ratio needs slowing down in the ratio 1 to 2.54.

ORIGIN OF 127 GEAR

If we want a reduction of 1 to 2.54 in gear form, then these quantities will have to be multiplied up to eliminate the 0.54 decimal portion. Multiplying by 100 gives gears of 100 and 254 teeth and on dividing each by 2 we have wheels of 50 and 127 teeth. And it is in this way that the seemingly strange 127 gear arrives on the scene. Further, when you consider that a lathe geared for cutting an English thread of so many to the inch can be made to cut that number of threads/CM by the inclusion of the 50-127 reduction gearing, you will see why the 50-127 gears are sometimes referred to as 'translation gears'. The 50 gear, however, does not appear in all gear trains because it often cancels with another of the gears. For example, taking our original 8/20 ration for 20 tpi, and introducing the 50/127 reduction, we have 8/20 × 50/127 = 20/127.

The 127 wheel holds a prime number of teeth, and is therefore the smallest that will offer an exact translation ratio. As will be seen, when slight pitch errors can be permitted or ignored − as is usually the case — approximations for the 50/127 ratio can be arranged with gears holding considerably less teeth.

Whilst the foregoing analysis serves as an introductory explanation, ordinarily metric pitches are invariably expressed in millimetres, and the term thrds/CM is not recognised: neither is it necessary to change a metric pitch into a thrds/CM value as an individual operation when calculating gear trains — the necessary figures are written into the formula which is evolved from the following reasoning —

$$\text{Since threads/CM} = \frac{10}{\text{Pitch in mm}}$$

we may initially position 10/P in place of tpi in the standard quadrant gearing formula for threads per inch with an English leadscrew (gearing = LSTPI/tpi to be cut) and we have:

$$\frac{\text{Drivers}}{\text{Driven}} = \frac{\text{LSTPI}}{\frac{10}{P}}$$

we then introduce the necessary 1 to 2.54 reduction by multiplying by 50/127 which gives the general formula:

$$\frac{\text{Drivers}}{\text{Driven}} = \frac{\text{LSTPI}}{1} \times \frac{P}{10} \times \frac{50}{127}$$

This formula, where LSTPI = tpi of leadscrew, and P = the metric pitch to be cut, expressed in mm, is applicable to leadscrews of *any* number of threads/inch, but since in practice we are hardly likely to encounter English leadscrews with other than 4, 5, 6, 8 or 10 tpi, the formula can be resolved into five simple and basic examples, any one of which is easily committed to memory, although we will probably only encounter Imperial leadscrews of 4 and 8 tpi today.

LEADSCREW TPI	FORMULA
4	$\dfrac{\text{Drivers}}{\text{Driven}} = \dfrac{2P}{5} \times \dfrac{50}{127}$
5	$\dfrac{\text{Drivers}}{\text{Driven}} = \dfrac{P}{2} \times \dfrac{50}{127}$
6	$\dfrac{\text{Drivers}}{\text{Driven}} = \dfrac{3P}{5} \times \dfrac{50}{127}$
8	$\dfrac{\text{Drivers}}{\text{Driven}} = \dfrac{4P}{5} \times \dfrac{50}{127}$
10	$\dfrac{\text{Drivers}}{\text{Driven}} = \dfrac{P}{1} \times \dfrac{50}{127}$

In each example, P represents the metric pitch (in mm) to be cut. The reason why these five examples are not reduced to their lowest terms (e.g. 20P/127 for a leadscrew of 4 tpi) is that each formula leaves the translation ratio of 50/127 intact, and later, when we discuss approximations for the ratio 50/127, the above formula will be used with the symbol *T* in place of 50/127, and the *T* will represent any one of those approximations.

Note: the 50/127 translation ratio is seldom used today, even by some of the best industrial lathes. However, these notes would be incomplete without a few examples showing the 50/127 translator in use.

USE OF FORMULAS

A leadscrew of 4 tpi would most probably be driven through a selective gearbox, thus affording the student no opportunity to set his own gears on a quadrant. Accordingly we will confine our attention to simple lathes with leadscrews of 8 and 10 tpi.

Example 1. To gear a leadscrew of 8 tpi for a metric pitch of 1.5 mm.

Substituting for P in the formula, we have:

$$\frac{\text{Drivers}}{\text{Driven}} = \frac{4 \times 1.5}{5} \times \frac{50}{127}$$

$$= \frac{6}{5} \times \frac{50}{127} = \frac{60}{127}$$

showing we require a 60 wheel as a driver and the 127 wheel as a driven, or lead-screw gear.

Example 2. To gear a leadscrew of 8 tpi for a metric pitch of 0.75 mm.

Substituting for P we have:

$$\frac{\text{Drivers}}{\text{Driven}} = \frac{4 \times 0.75}{5} \times \frac{50}{127} = \frac{30}{127}$$

showing we require a 30 wheel as a driver, and the 127 as a driven, or lead-screw gear.

Example 3. To gear a leadscrew of 8 tpi for a metric pitch of 0.6 mm.

Substituting for P, we have:

$$\frac{\text{Drivers}}{\text{Driven}} = \frac{4 \times 0.6}{5} \times \frac{50}{127}$$

$$= \frac{2.4}{5} \times \frac{50}{127} = \frac{24}{127}$$

However, if our change gears rise by fives, a 24 will not be available and we must split the 24 into, say, 4 × 6 and write:

$$\frac{4}{1} \times \frac{6}{127}$$

we may then multiply the 6 and 1 by 5, giving:

$$\frac{4}{5} \times \frac{30}{127}$$

and bringing the 4 and 5 to gear sizes by multiplying by ten we have:

$$\frac{\text{Drivers}}{\text{Driven}} = \frac{40}{50} \times \frac{30}{127}$$

which could be set on the quadrant:

$$40 - \text{A} - 50$$
$$30 - 127$$

The basic formula for dealing with a leadscrew of 10 tpi is of course similarly dealt with. For convenience the formula is here repeated:

$$\frac{\text{Drivers}}{\text{Driven}} = \frac{P}{1} \times \frac{50}{127}$$

where P is the pitch to be cut, in mm.

Example 1. The gear a leadscrew of 10 tpi for a metric pitch of 0.75 mm.

Substituting for P we have:

$$\frac{0.75}{1} \times \frac{50}{127}$$

multiplying 0.75 and 1 by 4 to eliminate the decimal point we have:

$$\frac{3}{4} \times \frac{50}{127} = \frac{30}{40} \times \frac{50}{127}$$

which might set on the quadrant:

$$30 - \text{A} - 40$$
$$50 - 127$$

Example 2. To gear a leadscrew of 10tpi for a metric pitch of 3.0mm.

Substituting for P we have:

$$\frac{3}{1} \times \frac{50}{127}$$

and multiplying 3 and 1 by 20 to bring to gear sizes we have:

$$\frac{60}{20} \times \frac{50}{127}$$

where it would probably be convenient to adopt the 50 as a first driver and set on the quadrant:

$$50 - \text{A} - 20$$
$$60 - 127$$

However, in this example the student should note the 50-20 step-up ratio, which although followed by an approximate 1-2 reduction (60/127) would place undesirable loading on the first part of the gear train. In this example, too, the overall step-up is only 150-127, or about 1 to 1.18 to 1 and obviously, but for the large number of teeth on the 127 gear, the initial 50-20 step-up would be unnecessary. When we deal with the question of approximations for the ratio 50/127 we will see how such awkward step-up ratios can generally be avoided. Admitted, many text-books unhesitatingly show large driving wheels – even up to 80 or more teeth driving a 20 for example, but in these instances it appears that writers get carried away by pure theory, and overlook the stresses imposed on such gear trains, when they are put to practical use.

GEARS OF REDUCED PITCH

The change gears supplied with the lathe illustrated in Fig. 10 are of No. 20 DP, and for all ordinary purposes may be assembled into fairly compact gear trains, but a gear of 127 teeth in the same diametral pitch has an outside or whole diameter of 6.45 inches (approximately 164 mm) giving a somewhat cumbersome wheel over which the change gear cover will not fit. Such a large gear also makes the setting for some metric pitches exceedingly voluminous: indeed many calculated settings are impossible. With these limitations in mind and a wish to make practical tests for these notes, it seemed worth investigating the possibilities of using a 127 gear of reduced diametral pitch (smaller teeth) which could nevertheless be assembled in gear trains initially driven by gears of the diametral pitch standard to the lathe. The photograph, Fig. 10A shows the reduced DP 127 gear in use, with the quadrant set for cutting a thread of 2.0 mm pitch.

For the smaller 127 gear (and mating gears) a DP of No. 30 was chosen, and by this means the diameter of the original 127 gear was reduced to a comfortable 4.4 inches (approximately 109 mm) and with standard change gears and a reduced pitch gear of 40 teeth to mesh with the new 127 gear, metric gear trains can be readily assembled in the manner shown in the right-hand table, No. T4, where the reduced DP gears are shown in italics.

CHECKING METRIC GEAR TRAINS

When a gear train has been calculated for cutting a metric thread pitch from an English leadscrew it is advisable to check that the computed gearing will in fact cut the metric pitch required, and for this purpose we may note that the pitch in mm

Fig. 10. The author's work-horse. A Myford lathe with quick acting rack tailstock, instant change to or from backgear, special clutch giving repeat pick up for all thread pitches, independent self-act reduction giving instant change to or from screwcutting ratios, adjustable carriage dead stop — also serves as a thread run out stop — front locking change gear studs and quick lock for quadrant. Centre height increased to 3⅝ in. (about 92.0 mm.).

Fig. 10A. Showing a 127T 'translation' gear in use. The 127 here is of No. 30 DP, with a meshing 40T, the remaining gears are of No. 20 DP: the Myford standard. The quadrant is set for a thread of 2.0 mm. pitch from a leadscrew of 8 tpi. See also text.

The pinion extreme lower left drives any gear train used for screwcutting when brought into mesh by lowering the quadrant. This pinion rotates at about 1/10 lathe spindle speed, thus self-act feed rates are always available.

from *any* gearing (including approximation gearing, q.v.) used in conjunction with an English leadscrew may be found from the formula:

$$\text{Pitch in mm} = \frac{\text{Drivers}}{\text{Driven}} \times \frac{25.4}{\text{Leadscrew TPI}}$$

Thus, for the five leadscrews set out on page 41 we find the checking formula resolves to:

LEADSCREW TPI	CHECKING FORMULA
4	$P = \dfrac{\text{Drivers}}{\text{Driven}} \times \dfrac{127}{20}$
5	$P = \dfrac{\text{Drivers}}{\text{Driven}} \times \dfrac{127}{25}$
6	$P = \dfrac{\text{Drivers}}{\text{Driven}} \times \dfrac{127}{30}$
8	$P = \dfrac{\text{Drivers}}{\text{Driven}} \times \dfrac{127}{40}$
10	$P = \dfrac{\text{Drivers}}{\text{Driven}} \times \dfrac{127}{50}$

By way of example we will now check the gearing previously calculated for a selection of metric pitches from leadscrews of 8 and 10 tpi:

Example No. 1: 0.6 mm pitch from a leadscrew of 8 tpi was calculated as requiring 40 and 30 drivers, and 50 and 127 driven, hence:

$$\frac{40}{50} \times \frac{30}{127} \times \frac{127}{40} = 0.6 \text{ mm}$$

Example 2. 0.75 mm pitch from a leadscrew of 10 tpi was found to require 30 and 50 drivers and 40 and 127 driven gears, hence:

$$\frac{30}{40} \times \frac{50}{127} \times \frac{127}{50} = 0.75 \text{ mm}$$

Example 3. The quadrant of a lathe with a leadscrew of 8 tpi is found to be set thus:

$$40 - 65 - 40$$
$$32 - 29$$

What thread pitch or pitches would this gearing give if used for screwcutting?

TABLE T4

PITCH MM	QUADRANT GEARING A.	QUADRANT GEARING B.
3·5	50 — 25 / 70 — 127	50 — 25 / 70 — 40 *40 — 127*
3·0	50 — 25 / 60 — 127	50 — 25 / 60 — 40 *40 — 127*
2·5	50 — 20 / 40 — 127	50 — A — 20 / 40 — 127
2·0	40 — 25 / 50 — 127	50 — A — 25 / 40 — 127
1·75	50 — 25 / 35 — 127	35 — A — 20 / 40 — 127
1·5	50 — 25 / 30 — 127	30 — A — 20 / 40 — 127
1.25	50 — A — 127	25 — A — 20 / 40 — 127
1·0	40 — A — 127	40 — A — 40 / 40 — 127
0·8	40 — 25 / 20 — 127	40 — A — 50 / 40 — 127
0·75	30 — A — 127	30 — A — 40 / 40 — 127
7·0	35 — 50 / 40 — 127	35 — A — 50 / 40 — 127
0·6	40 — 50 / 30 — 127	30 — A — 50 / 40 — 127
0·5	20 — A — 127	20 — A — 40 / 40 — 127
0·45	20 — 50 / 45 — 127	30 — 60 / 45 — 50 *40 — 127*
0·4	20 — 50 / 40 — 127	20 — A — 50 / 40 — 127

Table T4. Gearing for metric pitches with a leadscrew of 8 tpi and the exact 50/127 translation ratio. Gearing at left is with standard No. 20 DP gears, gearing at right is with a 40 and 127 gear of reduced DP. See also text. Mixing of reduced D.P. gears (shown italicised) offers a more compact assembly. Both systems give pitches to an accuracy equal to that of the leadscrew.

(a) 40 — 65 — 40 is merely a 1-1 ratio, so can be ignored. The effective ratio is 32-29. The metric pitch given would therefore = 32/29 x 127/40 = 3.503448 mm i.e. 3.5 mm with a pitch error of plus 0.003448 mm.

(b) The inch pitch given by the gearing = 32/29 × 1/8 = 4/29 = 0.137931 in. (exactly $7\frac{1}{4}$ tpi).

ALTERNATIVE TRANSLATION GEARING

(Approximations for the exact value 50/127)

Although use of the 50/127 ratio offers theoretically error-free Imperial-metric conversions, obviously the pitch accuracy of any thread cut by means of a leadscrew cannot exceed the pitch accuracy of the leadscrew itself, and, as already hinted in Section 2, lathe leadscrews, although sufficiently accurate for general run-of-the-mill threading, would be most unlikely to hold an accuracy similar to that required of jig-borer feed-screws, for example, where the error has to be less than plus or minus one tenth of a thousandth of an inch per 16 inch length (0.0025 mm in 406 mm) — see "Holes, Contours and Surfaces" by Richard F. Moore & Frederick C. Victory. Published by The Moore Special Tool Company, Bridgeport, Connecticut.

Moreover, lathe leadscrews are generally exposed to swarf and grit arising from machining operations, consequently a lathe leadscrew of supreme accuracy could not be expected to remain in that condition for any worthwhile period.

The foregoing facts coupled with the awkward size of gears with 127 teeth led to the adoption of approximations for the ratio 50/127, and perhaps the most commonly quoted approximation is 63/160 which appears to have been evolved circa 1914 (See "Fowler's Mechanics & Machinists Pocket Book and Diary" — Sixth Annual Edition, 1914, Edited by William H. Fowler, page 281). Now, although 63/160 may appear even more cumbersome than 50/127, it can in fact be broken down into elements: 7/4 × 9/40, 3/4 × 21/40 or 7/8 × 9/20 etc., thus considerably reducing gear sizes and the total number of teeth in some gear trains. For example, gearing for a pitch of 4.0 mm with a leadscrew of 8 tpi and the 63/160 translator is reduced to:

$$35 — A — 25$$
$$45 — 50$$

wherein, neglecting the number of teeth in the 'A' wheel, the teeth total = 155, and no effective gear is required with more than 50 teeth. The same 4.0 mm pitch using the 50/127 translator requires gearing:

$$40 — A — 20$$
$$80 — 127$$

showing a teeth total of 267, quite apart from the undesirability of the 40-20 step-up followed by an 80 driver.

If a 63 wheel is available, gearing for a 4.0 mm pitch with a leadscrew of 8 tpi becomes 63/50, thus offering a further reduction of the teeth total to 113, neglecting any necessary idle gears. However, extensive investigations by the author show that in general a 63T gear is not of sufficient use to warrant the addition of one to a set of change gears.

Occasionally a 63 wheel may be found amongst change gears with a second-hand lathe, and this sometimes leads to the belief that because 63 is approximately equal to 127/2, the ratio 25/63 should be used in place of, and as an approximation for, the ratio 50/127. Now, of course, while there can be no 'law' against using 25/63, this ratio is one of the poorest possible approximations for

50/127, the error being plus 7.9 thou. inch per inch, or plus one part in 126 parts, whereas when 63 (or 7 × 9) is used as a *driving* wheel – as the originator no doubt intended – the error is 63 times *less*, i.e. plus one thousandth of an inch in 8 inches, or plus one mm in 8000 mm if you prefer to look at it in that way.

It will be seen however that in converting from metric to English – i.e. cutting tpi from a metric leadscrew – translation ratios have to be inverted, so under these conditions a 63 would properly become a driven gear, although even here, a 63T wheel as such is not remarkably useful. 63 is also a driven gear in approximations No. 13 and 16, (Table T5) before inversion, but we can be fairly sure that Nos.13 and 16 were unknown to the originator of the 63/160 approximation.

LIST OF APPROXIMATIONS

The foregoing notes on the 63/160 approximation will have formed a useful introduction to the subject in general. Investigations by the author revealed well over 100 approximations for the ratio 50/127 and many more could be calculated. So far, the range extends from 21/50 with an error of minus one part in 255 parts (one inch in 7 yards) to 3/5 × 43/44 × 47/70 with an error of plus one inch in 12 miles. The Table T5 sets out 46 of these approximations, those selected being based upon potential value (for use in selective gearboxes, for example) and actual value for use in quadrant setting. Except where otherwise stated, the approximations were newly calculated by the author, the method of calculation being as follows:

(1). Multiply 127 by a number N, and let the product $= M$ Note: N may $=$ any number the prime factors of which should not exceed the number of teeth in the largest change gear that it would be con-

venient to use, say 53.

(2). Increase or reduce M (in turn) by 1, 2 & 3, thus forming a new number M'

(3). Note prime factors on numerator, and (where possible) factorise denominators offered by:

$$\frac{N}{M'} \times \frac{50}{1}$$

But: reject any quotients in M' showing prime factors in excess of the number of teeth that it would be convenient to use in gear form, say 73.

Example 1. Let $N = 63$ (7 × 9)
$$127 \times 63 = 8001 = M$$
$$M \text{ minus } 1 = 8000 = M'$$

The approximation will now $= \dfrac{N}{M'} \times \dfrac{50}{1}$

$$= \frac{7 \times 9 \times 50}{8000} = \frac{7}{4} \times \frac{9}{40} \begin{array}{l} \text{(No. 15 in} \\ \text{Table)} \end{array}$$

Example 1A. Let $N = 63$ (7 × 9)
$$127 \times 63 = 8001 = M$$
$$M \text{ plus } 1 \ = 8002 = M'$$

The approximation $= \dfrac{N}{M'} \times \dfrac{50}{1}$

$$= \frac{7 \times 9 \times 25}{4001}$$

But, as 4001 is a prime and cannot therefore be factorised into gear sizes, this approximation must be rejected

Example 2. Let $N = 62$ (2 × 31)
$$127 \times 62 = 7874 = M$$
$$M \text{ plus } 1 = 7875 = M'$$

The approximation $= \dfrac{N}{M'} \times \dfrac{50}{1}$

$$= \frac{2 \times 31 \times 50}{7875}$$

$$= \frac{4}{5} \times \frac{31}{63}\text{(No.16 in Table)}$$

TABLE T5 LIST OF APPROXIMATIONS FOR 50/127

No.	Approximation	Error	Remarks	Omits
1.	$\frac{13}{17} \times \frac{19}{25} \times \frac{21}{31}$	−1 in 658750	Contains Primes 3-5-7-13-17-19-31	Omits 2 & 11
2.	$\frac{5}{16} \times \frac{31}{29} \times \frac{33}{28}$	+1 in 129920	Credit Mr. F. Butler	
3.	$\frac{19}{30} \times \frac{23}{37}$	−1 in 55500		
4.	$\frac{5}{7} \times \frac{13}{19} \times \frac{29}{36}$	−1 in 47880	Contains Primes 2-3-5-7-13-19-29	Omits 11
5.	$\frac{6}{7} \times \frac{8}{11} \times \frac{12}{19}$	+1 in 36575	Contains Primes Patented by Colchester Lathes 2-3-7-11-19	Omits 13
6.	$\frac{7}{27} \times \frac{41}{27}$	−1 in 36450		
8.	$\frac{17}{31} \times \frac{28}{39}$	+1 in 30225		
8.	$\frac{41}{29} \times \frac{22}{79}$	+1 in 28637		
9.	$\frac{19}{34} \times \frac{31}{44}$	+1 in 24933		
10.	$\frac{11}{19} \times \frac{17}{25}$	−1 in 23750		
11.	$\frac{2}{9} \times \frac{19}{13} \times \frac{40}{33}$	−1 in 19305	Contains Primes 2-3-5-11-13-19	Omits 7
12.	$\frac{2}{7} \times \frac{13}{7} \times \frac{23}{31}$	−1 in 18987		
13.	$\frac{5}{6} \times \frac{25}{21} \times \frac{25}{63}$	−1 in 15876		
14.	$\frac{3}{13} \times \frac{29}{17}$	−1 in 11050		
15.	$\frac{7}{4} \times \frac{9}{40}$	+1 in 8000	Traditional (63/160)	

No.	Approximation	Error	Remarks
31.	$\frac{24}{61}$	−1 in 1525	
32.	$\frac{20}{31} \times \frac{25}{41}$	−1 in 1271	
33.	$\frac{10}{27} \times \frac{50}{47}$	+1 in 1269	
34.	$\frac{9}{22} \times \frac{25}{26}$	−1 in 1144	
35.	$\frac{1}{3} \times \frac{59}{50}$	−1 in 1070	
36.	$\frac{8}{7} \times \frac{10}{29}$	+1 in 1050	
37.	$\frac{7}{12} \times \frac{25}{37}$	+1 in 888	
38.	$\frac{5}{16} \times \frac{39}{31}$	−1 in 707	
39.	$\frac{5}{6} \times \frac{25}{53}$	−1 in 636	
40.	$\frac{5}{16} \times \frac{53}{42}$	+1 in 610	
41.	$\frac{8}{13} \times \frac{25}{39}$	+1 in 507	
42.	$\frac{28}{71}$	+1 in 592	
43.	$\frac{11}{28}$	−1 in 465	Traditional used by author
44.	$\frac{15}{38}$	+1 in 380	Traditional used by author
45.	$\frac{4}{9} \times \frac{8}{9}$	1 in 290	
46.	$\frac{20}{51}$ or $\frac{4}{3} \times \frac{5}{17}$	−1 in 252	

No.	Approximation	Error	Note
16.	$\frac{4}{5} \times \frac{31}{63}$	−1 in 7875	
17.	$\frac{5}{14} \times \frac{43}{39}$	+1 in 5460	Used extensively by the author
18.	$\frac{1}{2} \times \frac{37}{47}$	−1 in 4700	
19.	$\frac{25}{39} \times \frac{35}{57}$	−1 in 4446	
20.	$\frac{5}{7} \times \frac{27}{49}$	−1 in 3430	
21.	$\frac{5}{4} \times \frac{23}{73}$	+1 in 2920	Originally used by Drummond Lathes
22.	$\frac{10}{13} \times \frac{22}{43}$	−1 in 2795	
23.	$\frac{22}{49} \times \frac{50}{57}$	+1 in 2793	
24.	$\frac{21}{29} \times \frac{25}{46}$	−1 in 2668	
25.	$\frac{21}{31} \times \frac{25}{43}$	+1 in 2666	
26.	$\frac{20}{33} \times \frac{50}{77}$	−1 in 2541	
27.	$\frac{5}{8} \times \frac{17}{27}$	−1 in 2160	
28.	$\frac{15}{17} \times \frac{25}{56}$	+1 in 1904	Traditional. Used by author.
29.	$\frac{13}{33}$	+1 in 1650	
30.	$\frac{13}{14} \times \frac{25}{59}$	−1 in 1652	

Example 2A Let $N = 62 \ (2 \times 31)$
$$127 \times 62 = 7874 = M$$
$$M \text{ minus } 2 = 7872 = M'$$

The approximation $= \dfrac{2 \times 31 \times 50}{7872}$

$$= \frac{25}{41} \times \frac{31}{48}$$

(Not considered of sufficient value to include in Table).

Example 3. Let $N = 3 \times 7 \times 13 \times 19$
$$= 5187$$
$$127 \times 5187 = 658749 = M$$
$$M \text{ plus } 1 = 658750 = M'$$

Approx $= \dfrac{N}{M'} \times \dfrac{50}{1}$

$$= \frac{3 \times 7 \times 13 \times 19 \times 50}{658750}$$

$$= \frac{3 \times 7 \times 13 \times 19}{13175}$$

and, by trial, 13175 factorises into reasonably low primes:
$5 \times 5 \times 17 \times 31$ whereupon, re-forming into a three-element approximation we have:

$$\frac{13}{17} \times \frac{19}{25} \times \frac{21}{31} \quad \text{(No. 1 in Table)}$$

When computing approximations for the value 50/127 we may note that (1) the higher the N multiplier, and the less the deduction to form M' the smaller the error in the approximation. (2) The suitability of any N multiplier is purely a matter of arbitrary selection and trial. (3) Trial factorisation of the M' denominators obviously need not be continued beyond trial division by prime 73, or the number of teeth held by a largest gear considered to be of reasonable size.

49

NOTES ON THE TABLE OF APPROXIMATIONS. T5

The Table is presented in descending order of accuracy.

The error in any approximation is easily revealed by multiplying the approximation in question by 127/50, cancelling where possible, e.g. No.11

$$\frac{2}{9} \times \frac{19}{13} \times \frac{40}{33} \text{ and } \times \frac{127}{50} = \frac{19304}{19305}$$

Here, as the denominator 19305 exceeds the numerator by 1 (one), the error is minus 1 (one part) in 19305 parts.

Example 2. To find the error in No. 21 approximation:

$$\frac{5}{4} \times \frac{23}{73} \times \frac{127}{50} = \frac{2921}{2920}$$

Hence, as the numerator exceeds the denominator by 1 (one), the error is plus 1 (one) in 2920. This approach, incidentally, is thought to be more revealing than errors expressed in percentages.

USE OF APPROXIMATIONS

Hitherto it appears from a study of various text-books on lathe screwcutting that gearing for metric pitches to be cut from an English leadscrew (and vice versa) was, or is, more generally based upon or evolved from the use of only one approximate translation ratio, perhaps No.15, 63/160, but in the course of lathe screwcutting over seven hundred feet of various threads, (individually seldom exceeding about 2 in. in length) both English and metric, from 4.0 mm pitch to 1.5 mm pitch and from 8 tpi to 26 tpi., the procedure now favoured by the author (when translations are necessary) is to select an approximate translation ratio that:

(1). Offers the simplest gear train, i.e. avoids double compound gear trains.

(2). gives the most convenient ratio for repeat pick-up. (Avoidance of crossed threading: see Section 5)

(3). And to generally ignore the error-value of the approximation, although of course, one would choose the best approximation, consistent with conditions (1) and (2).

This approach can save a considerable amount of time.

Some of the higher value (minimum error) approximations (Nos. 1 to 14) will not necessarily be of use for individual quadrant settings, but Nos. 1, 4 and 11 may be of value for incorporation in selective gearboxes for the reasons that:

(1) These approximations are of minimum error.

(2) Each hold nearly all the primes needed for 'threads per inch' gearing, and would thus avoid duplication of those primes elsewhere in a gearbox.

CALCULATIONS.

Although any individual reader will probably have to deal with, at most, two English leadscrews from the list of five on page 41 of this Section, for convenience, all five are here repeated, with symbol T in place of the exact 50/127 translation ratio, the T now representing any one of the 46 approximation ratios in Table T5

LEADSCREW TPI	FORMULA
4	$\frac{Drivers}{Driven} = \frac{2P}{5} \times T$
5	$\frac{Drivers}{Driven} = \frac{P}{2} \times T$
6	$\frac{Drivers}{Driven} = \frac{3P}{5} \times T$
8	$\frac{Drivers}{Driven} = \frac{4P}{5} \times T$
10	$\frac{Drivers}{Driven} = \frac{P}{1} \times T$

PRACTICAL EXAMPLES

Some examples from the author's experience will now be given.

Example 1. To gear a leadscrew of 8 tpi for 1.5 mm pitch.

This 1.5 mm pitch is sometimes used to concentrically thread nuts which, at the same chucking, are subsequently finish sized with a ground-thread tap. Consequently pitch accuracy during lathe screwcutting is unimportant and the 15/38 translator (No.44 in Table) can be used.

The necessary gearing for a leadscrew of 8 tpi is:

$$\frac{\text{Drivers}}{\text{Driven}} = \frac{4P}{5} \times T$$

Substituting for *P* and *T* we have:

$$\frac{\text{Drivers}}{\text{Driven}} = \frac{4 \times 1.5}{5} \times \frac{15}{38} = \frac{18}{38}$$

But, as a 40T first driver is both necessary and convenient (to suit the limitations imposed by a special clutch – see section 5). the 18/38 is rearranged to 2/1 × 9/38. The 9 and 1 are then multiplied by 4, giving 2/4 × 36/38, and the 2 and 4 multiplied by 20, giving a final ratio of 40/80 × 36/38, which sets on the quadrant:

$$40— \text{A} —38$$
$$36—80$$

This 1.5 mm pitch is often followed (or preceded) by call for a thread of 14 tpi for which a 70T wheel is substituted for the 80 and meshed with the 38 which then becomes a second idle wheel, and the 36 is merely left in position with nothing to do:

$$40— \text{A} —38—70$$
$$36$$

and of course 40-70 gears a leadscrew of 8 tpi for 14 tpi.

Example 2. To gear a leadscrew of 8 tpi for 4.0 mm pitch.

Using the formula 4P/5 × T, and select-

ing No. 17 translator we have:

$$\frac{\text{Drivers}}{\text{Driven}} = \frac{4 \times 4}{5} \times \frac{5}{14} \times \frac{43}{39} = \frac{8}{7} \times \frac{43}{39}$$

and multiplying 8 and 7 by 5 we have 40/35 × 43/39 which conveniently sets on the quadrant:

$$40— \text{A} —39$$
$$43—35$$

As with example No. 1, this 4.0 mm pitch is often preceded or followed by gearing for a thread of 8 tpi (and 14 tpi) and this is quickly set by substituting a 40 for the 35 and meshing the 40 with the 39 which now becomes a second idle gear:

$$\frac{40— \text{A} —39—40}{43} = 8 \text{ tpi, and:}$$

Similarly

$$\frac{40— \text{A} —39—70}{43} = 14 \text{ tpi}$$

In general, and assuming any necessary gears are available and that pick-up does not require special consideration, when choosing an approximation for 50/127 from the table, gearing can be simplified by selecting an approximation wherein a fractional element of that approximation cancels wholly or partly with the first element of the general gearing formula.

Example 3. For example, the general formula for gearing for metric pitches with a leadscrew of 8 tpi is 4P/5 × T. Now suppose we wish to gear for a metric pitch of 2.5 mm. Substituting 2.5 for P, we have (4 × 2.5)/5 × T = 10/5 × T, accordingly if we can find an approximation containing an element 5/10, or 1/2, the gearing will be simplified. Scanning the table of approximations we find No. 18: 1/2 × 37/47, consequently we have:

$$\frac{\text{Drivers}}{\text{Driven}} = \frac{10}{5} \times \frac{1}{2} \times \frac{37}{47} = \frac{37}{47}$$

which would set on the quadrant:

37 — A — A — 47 or, with a 40
first driver:

40 — A — 40
37 — 47

And this gearing would cut a pitch of:

37/47 × 127/40 = 2.499468 mm
showing a pitch error of minus 0.000532
mm, or 21 millionths of an inch, assuming
a leadscrew of perfect lead.

Example 4. We may require a pitch of 3.5
mm from a leadscrew of 8 tpi. The basic
requirement will then be (4 × 3.5)/5 × T =
14/5 × T. Now obviously the 5/14 of No.
17 approximation would cancel nicely,
leaving us with a 43 driver and 39 driven,
but of course, if we have only gears 20-
20-75 by fives plus one 38, we have to
search for approximations for 50/127
containing those gears, or primes of those
gears. No. 20 approximation: 5/7 × 27/49
holds primes 3, 5 and 7, and may be
rewritten: 5/7 × 3/7 × 9/7, thus:

$$\frac{\text{Drivers}}{\text{Driven}} = \frac{14}{5} \times \frac{5}{7} \times \frac{3}{7} \times \frac{9}{7}$$

wherein two of the 5's cancel, one of the
7's cancel with 14, giving numerator 2,
and we have:

$$\frac{\text{Drivers}}{\text{Driven}} \quad \frac{6}{7} \quad \frac{9}{7} \quad \frac{60}{70} \quad \frac{45}{35}$$

which would set on the quadrant:

45 — A — 35
60 — 70

And if we try using the No.44 approximation we arrive at gearing:

30 — A — 25
35 — 38

Example 5. Occasionally the approximations may be used to compensate for lack
of a particular gear. For example,
ordinarily, to gear for a non-standard pitch
of 1.7 mm would require a wheel such as
34, holding prime 17, and with a leadscrew of 8 tpi, the basic gearing would be:

(4 × 1.7)/5 × T = (4 × 17)/50 × T, but on
scanning the table of approximations we
find that No. 28, 15/17 × 25/56 holds a
denominator 17 which will cancel:

$$\frac{\text{Drivers}}{\text{Driven}} = \frac{4 \times 17}{50} \times \frac{15}{17} \times \frac{25}{56} = \frac{3}{4} \times \frac{5}{7}$$

which after multiplying up to gear sizes
would set on the quadrant:

30 — A — 35
25 — 40

which gives a pitch of 1.7008928 mm,
showing a pitch error of plus 0.0008928
mm, or plus 35 millionths of an inch with
a leadscrew of perfect lead.

CHANGE GEAR TABLE

For the benefit of those possessing only
the more customary set of change gears
20-20-75 by fives, plus one 38, the Table
T6 overleaf offers gearing for metric
pitches from 4.0 mm to 0.25 mm with a
leadscrew of 8 tpi and requires no special
gears with the exception of 0.25 mm pitch
which requires an 80 wheel, and an
alternative train for 0.35 mm pitch, which
requires a 33 wheel. In a majority of
examples, alternative gear trains are
given, and seven different approximations
for the ratio 50/127 are used throughout
the table. We may note that although the
63/160 translation ratio is used for some
gearing, it is always used in the basic form
7/4 × 9/40 which of course eliminates
call for a 63 gear, which as a driving
wheel would be inconveniently large,
especially for the finer pitches, where, for
example, a pitch of 0.5 mm would call for
gearing 1/10 × 63/40 or 1/5 × 63/80.

Regrettably, because of the large speed
reductions required for the finer pitches
0.75 mm down to 0.25 mm, the double
compound gear trains are unavoidable.

The actual pitch given by any of the
gear trains in the table T6 may be ascer-

tained from the formula already given:

$$\text{Pitch (mm)} = \frac{\text{Drivers}}{\text{Driven}} \times \frac{127}{40}$$

For example, taking the first of the three gear trains given for a pitch of 2.0 mm we have 45, 35 and 30 drivers, and 25, 50 and 60 driven, hence:

$$\frac{45}{25} \times \frac{35}{50} \times \frac{30}{60} \times \frac{127}{40} = 2.00025 \text{ mm}$$

MYFORD'S APPROXIMATION

Myford Limited, the well-known lathe makers, invented a further modification of the 63/160 approximation ratio, and this takes the form 3/4 × 21/40 calling for only one special gear of 21 teeth, and when this is used in conjunction with a leadscrew of 8 tpi and their standard gears by fives, offers the following basic formula:

$$\frac{\text{Drivers}}{\text{Driven}} = \frac{3P}{5} \times \frac{21}{40}$$

where P is the pitch to be cut. This approach offers neat gearing for the finer pitches, e.g.:

1.0 mm pitch   30 — A — 50
                               21 — 40

2.0 mm pitch   30 — A — 25
                               21 — 40

3.0 mm pitch   45 — A — 25
                               21 — 40

However, gearing for a pitch of 3.5 mm proves rather cumbersome because it calls for a 60 driver, and a 3-1 step-up, and large ratio increases are always undesirable:

60 — 20   21 — 50
        35 — 40

Of course a more 'balanced' train can be obtained by use of a 42 wheel:

42 — 20   30 — 50
        35 — 40

But if we are going to introduce yet more special gears to meet every adversity, then attempts at the formation of rules or systems are easily defeated.

EXTRACTION OF AN APPROXIMATION RATIO FROM A GEAR TRAIN

Occasionally when English/metric gearing contains an approximation for the exact ratio 50/127 (as will be evident from the absence of a 127 gear) it can be useful to ascertain the approximation used for 50/127 in the gearing.

The translation ratio can be found from the following formula:

$$T = \frac{\text{Drivers}}{\text{Driven}} \times \frac{10}{P \times \text{LSTPI}}$$

where LSTPI = threads per inch of leadscrew, P = the NOMINAL pitch in mm (as distinct from the approximate pitch given by the gearing) and T the approximation to be found.

By way of example, and referring to the Table T6 wherein the first of the two sets of gearing for a pitch of 1.75 mm from a leadscrew of 8 tpi is given as 30/35 × 45/70, what approximate translation was used?

The approximation T

$$= \frac{30}{35} \times \frac{45}{70} \times \frac{10}{1.75} \times \frac{1}{8}$$

$$= \frac{5}{7} \times \frac{27}{49} \quad \text{(No.20 in Table T5)}$$

And multiplying by 127/50 to reveal the error we have:
5/7 × 27/49 × 127/50 = 3429/3430
showing an error of minus one part in 3430.

53

METRIC PITCH MM	QUADRANT GEARING	ERROR + OR −1 IN	No. OF TRANSLATOR USED
4·0	35 —— A —— 25 / 45 —— 50	+ 8000	15
	30 —— A —— 25 / 40 —— 38	+ 380	44
3·5	45 —— A —— 35 / 60 —— 70	− 3430	20
	30 —— A —— 25 / 35 —— 38	+ 380	44
3·0	40 —— A —— 50 / 65 —— 55	+ 1650	29
	30 —— A —— 38 / 60 —— 50	+ 380	44
2·75	40 —— A —— 60 / 65 —— 50	+ 1650	29
	65 —— A —— A —— 75	+ 1650	29
	30 —— A —— 38 / 55 —— 50	+ 380	44
2·5	35 —— A —— 40 / 45 —— 50	+ 8000	15
	30 —— A —— A —— 38	+ 380	44
2·0	45 —— 25 / 35 —— 30 / 50 —— 60	+ 8000	15
	45 —— A —— 55 / 50 —— 65	− 1144	34
	30 —— A —— 38 / 40 —— 50		
1·75	30 —— A —— 35 / 45 —— 70	− 3430	20
	30 —— A —— 38 / 35 —— 50	+ 380	44
1·5	40 —— 60 / 50 —— 65 ; 35 —— 38	− 4446	19
	20 —— A —— 50 / 65 —— 55	+ 1650	29
	30 —— A —— 38 / 45 —— 75	+ 380	44
1·25	35 —— 50 / 45 —— 40 ; 30 —— 60	+ 8000	15
	45 —— 55 / 50 —— 40 ; 25 —— 65	− 1144	34
	30 —— A —— 38 / 25 —— 50	+ 380	15

METRIC PITCH MM	QUADRANT GEARING			ERROR + OR − 1 IN	No. OF TRANSLATOR USED
1·0	35 — 50 / 45 — 40	30 — 75		+ 8000	15
	25 — A — 55 / 45 — 65			− 1144	34
	20 — A — 38 / 30 — 50			+ 380	44
0·8	20 — A — 55 / 45 — 65			− 1144	34
0·75	40 — 38 / 35 — 60	25 — 65		− 4446	19
	30 — 40 / 25 — 65	45 — 55		− 1144	34
0·7	30 — 50 / 45 — 35	20 — 70		− 3430	20
	20 — 70 / 55 — 75	40 — 38		− 2794	23
	20 — 40 / 30 — 75	55 — 50		− 466	43
0·6	35 — 38 / 25 — 65	40 — 75		− 4446	19
	20 — 40 / 30 — 65	45 — 55		− 1144	34
0·5	30 — 60 / 45 — 55	25 — 65		− 1144	34
0·45	35 — 38 / 25 — 50	20 — 65		− 4446	19
0·4	25 — 50 / 45 — 55	20 — 65		− 1144	34
0·35	55 — 38 / 20 — 70	20 — 75		− 2794	23
	30 — 40 / 33 — 60	20 — 75		− 466	43
0·3	35 — 38 / 25 — 65	20 — 75		− 4446	19
0·25	25 — 55 / 45 — 65	20 — 80		− 1144	34

TABLE T6

METRIC PITCHES FROM A LEADSCREW OF 8 TPI GEARS 20–20–75 BY FIVES PLUS ONE 38.

TABLE T7
APPROXIMATIONS FOR Pi × 50/127 = 1 to 0.8085071 Symbol P.

Approx. No.	Approx. For P	Error in Approximation (Pi = 3.1415926)	
1	$\frac{5}{4}$	+1 in	94
2	$\frac{16}{13}$	−1 in	203
3	$\frac{41}{33}$	+1 in	221
4	$\frac{36}{29}$	+1 in	273
5	$\frac{37}{30}$	−1 in	352
6	$\frac{31}{25}$	+1 in	392
7	$\frac{58}{47}$	−1 in	441
8	$\frac{88}{71}$	+1 in	477
9	$\frac{57}{46}$	+1 in	541
10	$\frac{21}{17}$	−1 in	796
11	$\frac{26}{21}$	+1 in	991
12	$\frac{99}{80}$	+1 in	1896
13	$\frac{68}{55}$	−1 in	2555
14	$\frac{47}{38}$	−1 in	230000

WORM THREADS FOR GEARS SIZED BY MODULE. ENGLISH LEADSCREW GEARING

Just as an engineer working by inch measure gains a knowledge of the relative dimensions of a gear and its teeth from the diametral pitch number, so does the metric system of module values indicate the relative proportions of gears dimensioned in millimetres and sized by module. Whereas No. 1 diametral pitch = Pi by inch measure = 3.1415926 inch, circular pitch, No. 1 module = Pi in millimetres, No. 2 module = 2 × Pi = 6.283 . . . mm, and 0.5 module = Pi/2 = 1.570 . . . mm circular pitch.

Before gearing can be calculated for lathe screwcutting it is of course necessary to ascertain the module value of the gear for which a worm is to be made to mesh.

Given a gear suspected to be of metric origin, the module number may be found by the following method:

Count the teeth and add 2

Measure the whole diameter in mm

$$\text{then Module} = \frac{\text{Dia. in mm}}{\text{Teeth plus 2}}$$

For example, a gear has 40 teeth, and the outside, or whole diameter is 52.5 mm

$$\text{Module} = \frac{52.5}{42} = 1.25 \text{ mm}$$

We may also note:

$$\text{Module} = \frac{\text{Pitch diameter}}{\text{Number of teeth}}$$

(Pitch diameter here not to be confused with whole or 'outside' diameter.)

Also, Module in mm

$$= \frac{1}{\text{Diametral pitch}} \times \frac{25.4}{1}$$

Module by inch measure

$$= \frac{1}{\text{Diametral pitch}}$$

Circular pitch (inch) $= \dfrac{Pi}{DP}$

Circular pitch (mm) from module (mm) $= Pi \times \text{module}$

Circular pitch (inch) from module (mm) $= \dfrac{Pi \times M}{25.4}$

The formula for gearing an English leadscrew to cut worms sized by module reads:

$$\frac{\text{Drivers}}{\text{Driven}} = \frac{M}{10} \times \frac{\text{LSTPI}}{1} \times \frac{50 \times Pi}{127}$$

where M = module number, LSTPI = threads/inch of leadscrew, and Pi = 3.1415926.

This formula, however, although theoretically correct, when transposed into gear trains, produces only cumbersome arrangements for which there may not be sufficient space to set on a quadrant, even when Pi is approximated to 22/7. This difficulty can be overcome by use of approximations for 50 × Pi/127 and 14 such approx. ratios are given in Table T7 which is presented in ascending order of accuracy. With use of symbol P for any approx. ratio the gearing formula then reads:

$$\frac{\text{Drivers}}{\text{Driven}} = \frac{M}{10} \times \frac{\text{LSTPI}}{1} \times P$$

which, for a leadscrew of 8 tpi reduces to

$$\frac{\text{Drivers}}{\text{Driven}} = \frac{4M}{5} \times P$$

Example 1. Suppose we wish to find gearing for No.1 module (circular pitch = 3.14159 mm) with a leadscrew of 8 tpi. If we are limited to change gears 'by fives' we will naturally seek a reasonable looking approx. (*P*) that can be resolved into those gears. No. 11, 26/21 should serve, as it holds primes 3, 7 and 13, and we have:

$$\frac{\text{Drivers}}{\text{Driven}} = \frac{4}{5} \times \frac{2}{3} \times \frac{13}{7}$$

which after multiplying up to change gear sizes offers a quadrant setting:

40— A —35
        65—75

The circular pitch given = Drivers/driven × 127/40
= 40/35 × 65/75 × 127/40
= 3.14476 mm
i.e. plus 0.00317 mm on true CP
(about 1/10 thou. in. too coarse, assuming a leadscrew of perfect lead)

Example 2. To gear a leadscrew of 8 tpi for No. 1.25 module. We may note that for 1.25 module with an 8 tpi leadscrew, the first portion of our formula, 4M/5, cancels to 1 (one) so 1.25 module will be given by any of the 14 ratios for which gears shown in Table T7 are available.

SECTION 4

Lathes with metric lead-screws

Although one may reasonably assume that a thread of 8 to the inch is the present standard for lathes of the light to medium duty instrument type, it is not yet possible to say which of a choice of a few similar metric pitches will become popular when the metric system is fully operational.

The nearest integral ISO metric pitch to a thread of 8 to the inch is 3.0 mm (8.4666 tpi) and it is understood that leadscrews of 3.0 mm pitch will be adopted by the makers of the Hardinge HLV-H High Precision Lathe if circumstances dictate that they have to change.

QUADRANT GEARING CALCULATIONS.
GEARING FOR METRIC THREADS – METRIC LEAD-SCREWS

Metric threads are invariably designated by pitch, and except for explanatory purposes, a reference to so many threads/CM is neither recognised nor called for. The change gear formula therefore differs slightly from that used for calculating gearing for threads/inch with an English leadscrew.

When the leadscrew and component to be threaded are both expressed in terms of millimetre pitch, the formula reads:

$$\frac{\text{Drivers}}{\text{Driven}} = \frac{\text{Pitch of screw to be cut (mm)}}{\text{Pitch of leadscrew (mm)}}$$

Example 1. Pitch required 2.5 mm, with a leadscrew of 3.0 mm pitch. (Change gears by fives)

$$\frac{\text{Drivers}}{\text{Driven}} = \frac{2.5}{3}\text{multiply by 10}$$

$$= \frac{25}{30}\frac{\text{Driver}}{\text{Driven}}$$

Example 2. Pitch required 0.75 mm, 3.0 mm pitch leadscrew:

$$\frac{\text{Drivers}}{\text{}} = \frac{0.75}{3}\text{multiply by 100}$$

$$= \frac{75}{300} = \frac{5}{10} \times \frac{15}{30} = \frac{1}{2} \times \frac{1}{2}$$

$$= \frac{25}{50} \times \frac{30}{60} \text{ or}$$

$$\frac{20}{40} \times \frac{25}{50}$$

which would set on a quadrant:

$$20 - A - 40$$
$$25 - 50$$

or, of course, with an 80 wheel available, quadrant setting would become:

$$20 - A - A - 80$$

58

Example 3. Pitch required 4.0 mm with a leadscrew of 3.5 mm pitch.

$$\frac{\text{Drivers}}{\text{Driven}} = \frac{4}{3.5} \text{ multiply by 10} = \frac{40}{35}$$

= a 40 driver and 35 driven.

Example 4. Pitch required 1.5 mm with a leadscrew of 3.5 mm pitch.

$$\frac{\text{Drivers}}{\text{Driven}} = \frac{1.5}{3.5} = \frac{15}{35} = \frac{30}{70}$$

= a 30 driver and a 70 driven.

METRIC – METRIC CHECKING

To check that any gear train will produce the metric pitch for which the calculation was made, we may note that:

$$\text{Pitch (mm)} = \frac{\text{Drivers}}{\text{Driven}} \times \frac{\text{LS pitch (mm)}}{1}$$

Checking Example 1

The gearing for a pitch of 0.75 mm with a leadscrew of 3.0 mm pitch was found to be 20/40 × 25/50, hence:

$$\text{Pitch given} = \frac{20}{40} \times \frac{25}{50} \times \frac{3}{1}$$

$$= \frac{3}{4} = 0.75 \text{ mm}$$

Checking Example 2

The gearing for a pitch of 1.5 mm with a leadscrew of 3.5 mm pitch was found to be 30/70, hence:

$$\text{Pitch given} = \frac{30}{70} \times \frac{35}{10} = \frac{3}{2} = 1.5 \text{ mm}$$

GEARING FOR THREADS/ INCH WITH A METRIC LEAD- SCREW

Just as with an English leadscrew we were able to convert threads/inch into threads/CM by introducing into the gear train a reduction in the ratio 1 to 2.54 or 50/127 (or by introducing approximations for the ratio 50/127) so we may gear a metric leadscrew to cut so many threads/ CM, and then by *increasing* the leadscrew speed relative to the workpiece in the ratio 2.54 to 1, or 127/50, convert the threads/CM into threads/inch.

For example, a leadscrew of 3.0 mm pitch geared 1 to 3 will cut a pitch of 1.0 mm (10 threads/CM) but if we now speed up the leadscrew in the ratio 2.54 to 1, or 127/50, the pitch cut will become (using the checking formula):

$$\text{Pitch (mm)} = \frac{1}{3} \times \frac{127}{50} \times \frac{3}{1} = 2.54 \text{ mm}$$

and, of course, 2.54 mm is exactly equal to 0.1 inch pitch = 10 threads/inch.

APPROXIMATION GEARING

With the usual objections to using so large a wheel as the 127 in a *driving* position in the gear trains, we again turn to the use of approximations for the 127/50 ratio. For these approximations, the Table T5 may be referred to, but it is important to note that when these approximations are to be used for metric-English conversions, the approximations must be inverted. No. 29, for example, 13/33 becomes 33/13. We should also note that on inversion, the errors in the approximations are reversed from plus to minus and vice versa. For example, the error with No. 29 approximation, 13/33 is plus one in 1650, but on inversion to 33/13, the error changes to minus one in 1651.

BASIC METRIC/ENGLISH FORMULA

It can be shown that the basic formula for gearing a metric leadscrew of *any* pitch for cutting threads/inch reads:

$$\frac{\text{Drivers}}{\text{Driven}} = \frac{1}{\text{TPI}} \times \frac{10}{\text{LSP}} \times \text{T}$$

where TPI symbolises the threads/inch to be cut, LSP, the pitch of the leadscrew in millimetres, and T = 127/50, or any approximation for that ratio.

WHAT IS THE BEST APPROXIMATION FOR 127/50?

This depends upon the pitch of the leadscrew and the change gears available. For a leadscrew of 3.0 mm pitch, the traditional 160/63 approximation ratio resolves to:

Drivers/Driven = 4/TPI × 10/3 × 40/63 or
    4/TPI × 10/3 × 5/7 × 8/9

either of which offer somewhat cumbersome gear trains with gears by fives, chiefly because denominator 3 will not cancel with any numerator in the approximation.

Careful investigation by the author shows that for threads/inch from a leadscrew of 3.0 mm pitch and gears by fives plus one 38, the most convenient approximations are Nos. 19, 29, 34, 43, 44 & 45 (from Table T5) and these approximations are used in the accompanying Table T8 showing gearing for threads/inch from 6 to 72.

As examples of the way in which Table T8 was calculated let us take the three sets of gearing for a thread of 11 tpi from a leadscrew of 3.0 mm pitch.

We commence with the general formula:

$$\frac{\text{Drivers}}{\text{Driven}} = \frac{1}{\text{TPI}} \times \frac{10}{\text{LSP}} \times T$$

Then (1) Substituting the known figures,we have:

$$\frac{\text{Drivers}}{\text{Driven}} = \frac{1}{11} \times \frac{10}{3} \times T$$

Now, if we can find an approximation for T which (A) contains no primes outside those held in our gears by fives plus one 38, and (B) after inversion holds 11 or 3 or both in the numerator, then the gearing will be simplified. No. 29 approximation

after inversion to 33/13 looks promising, hence:

$$\frac{\text{Drivers}}{\text{Driven}} = \frac{1}{11} \times \frac{10}{3} \times \frac{33}{13} = \frac{10}{13}$$
$$= \frac{50\ \text{driver}}{65\ \text{driven}}$$

(2). Are there any other approximations with characteristics similar to those just outlined?

Scanning the Table T5 we find No. 34, which after inversion offers 22/9 × 26/25, hence:

$$\frac{\text{Drivers}}{\text{Driven}} = \frac{1}{11} \times \frac{10}{3} \times \frac{22}{9} \times \frac{26}{25}$$

which reduces to 8/9 × 13/15 and when multiplied up to gear size offers:

$$\frac{\text{Drivers}}{\text{Driven}} = \frac{40}{45} \times \frac{65}{75}$$

(In examples similar to this, not having any smaller multiple of prime 13 than 65 can be a nuisance when the 65 has to be used in a driving position).

(3). Lastly, approximation No. 44 is generally worth trying regardless of other considerations. No. 44 after inversion = 38/15, and we have:

Drivers/Driven = 1/11 × 10/3 × 38/15

which resolves to:

$$\frac{\text{Drivers}}{\text{Driven}} = \frac{50}{55} \times \frac{38}{45}$$

MINIMUM ERROR

It is interesting to note that 11 threads/inch from a leadscrew of 3.0 mm pitch could be arranged with No. 11 approximation which after inversion reads:

$$9/2 \times 13/19 \times 33/40$$

which, when included in the general formula reduces to:

$$\frac{\text{Drivers}}{\text{Driven}} = \frac{45}{40} \times \frac{26}{38}$$

and, assuming a 26 gear is available, would set on the quadrant:

$$26 — A — 38$$
$$45 — 40$$

The pitch error with this gearing and an accurate leadscrew would be plus about $3\frac{1}{4}$ inches a mile. However, without any multiple of prime 13 other than 65, the quadrant setting for the same ratio is somewhat clumsy:

$$65 — 40 \quad 20 — 50$$
$$45 — 38$$

and for this reason it was omitted from the Table T8.

METRIC/ENGLISH. CHECKING

The threads/inch given by *any* quadrant gearing used in conjunction with a metric leadscrew of *any* pitch may be obtained from the formula:

Threads/inch

$$= \frac{\text{Driven gears}}{\text{Driving gears}} \times \frac{127}{5 \times \text{LSP (mm)}}$$

(Note the inversion: driven/drivers)

For example (1). Taking the first of our three calculations for a thread of 11 tpi from a leadscrew of 3.0 mm pitch, we found that the gearing resolved to a 50 driver and a 65 driven, hence:

$$\text{Threads/inch} = \frac{65}{50} \times \frac{127}{15} = \frac{1651}{150}$$
$$= 11.006666$$

Note here that the 0.0066 ... decimal portion should not be read as 'of an inch' but as 'of a thread turn' hence the threads/inch will be 11 plus 6.6 thousandths of a thread turn.

If gearing is to be checked in terms of the pitch given by inch measure, the formula reads:

$$\text{Pitch (inches)} = \frac{\text{Drivers}}{\text{Driven}} \times \frac{5 \times \text{LSP (mm)}}{127}$$

hence the pitch given by the preceding example is:

$$\frac{50}{65} \times \frac{15}{127} = \frac{150}{1651} = 0.090854 \text{ inch}$$

The exact pitch of a thread of 11 tpi is 0.090909 inch, hence the pitch error in the gearing (introduced by use of the No. 29 approximation for 127/50) is minus 0.000055 inch.

Example 2. What (inch) pitch is produced from a leadscrew of 3.0 mm pitch by gearing 26/38 × 45/40?

26/38 × 45/40 × 15/127 = 0.0909137 inch, which, from the reciprocal = 10.999442 threads/inch, i.e. the thread will be too 'coarse- by 0.000558 of a thread turn, or plus 0.0000047 inch on pitch — assuming a leadscrew of perfect lead.

LEADSCREW OF 3.5 mm PITCH

The author's lathe was originally supplied with a leadscrew of 8 tpi, but when any significant number of metric threads are to be lathe screwcut, a metric leadscrew and half-nuts are substituted. On completion of the metric threading, the metric leadscrew is left in position until a sufficient number of English threads are required to warrant change back to the English leadscrew. This procedure offers the highest possible production rate for quantities of screws – 50 to 100 or so – in either language, although it does dictate that translation ratios be used when only a few threads in either language are required, and an adverse leadscrew is in position.

It may be thought that from a business point of view an additional lathe with a metric leadscrew would be installed. There are several reasons why this was not done. One is that the cost would not be justified by one working without staff. To justify the cost, any additional lathe

GEARING FOR THRDS/INCH FROM A LEADSCREW OF 30 MM PITCH GEARS 20–20–75 BY FIVES PLUS ONE 38.

TPI REQ	GEARING	ERROR +/– 1 IN	No. OF T RATIO USED
6	38 — A — 50 / 65 — 35	+ 4446	19
	50 — A — 30 / 55 — 65	– 1650	29
	38 — A — 30 / 50 — 45	– 380	44
7	50 — A — 35 / 55 — 65	– 1650	29
	40 — A — 30 / 50 — 55	+ 465	43
	38 — A — 35 / 50 — 45	– 380	44
8	50 — A — 40 / 55 — 65	– 1650	29
	38 — A — 40 / 50 — 45	– 380	44
9	50 — A — 45 / 55 — 65	– 1650	29
10	55 — A — A — 65	– 1650	29
	40 — A — 30 / 35 — 55	+ 465	43
	38 — A — A — 45	– 380	44
11	50 — A — A — 65	– 1650	29
	40 — A — 45 / 65 — 75	+ 1144	34
	38 — A — 45 / 50 — 55	– 380	44
12	38 — A — 50 / 65 — 70	+ 4446	19
	25 — A — 30 / 55 — 65	– 1650	29
	38 — A — 45 / 50 — 60	– 380	44
14	25 — A — 35 / 55 — 65	– 1650	29
	20 — A — 30 / 50 — 55	+ 465	43
16	25 — A — 40 / 55 — 65	– 1650	29
	25 — A — 30 / 35 — 55	+ 465	43
	25 — A — 40 / 38 — 45	– 380	44
18	25 — A — 45 / 55 — 65	– 1650	29

TABLE T8

TPI REQ	GEARING	ERROR +/- 1 IN	No. OF T RATIO USED
19	20 —— A —— 38 / 55 —— 65	1650	29
	20 —— A —— A —— 45	- 380	44
20	30 —— A —— 60 / 55 —— 65	- 1650	29
	20 —— A —— 30 / 35 —— 55	+ 465	43
22	25 —— A —— A —— 65	- 1650	29
	20 —— A —— 45 / 65 —— 75	+ 1144	34
	25 —— A —— 45 / 38 —— 55	- 380	44
24	25 —— A —— 60 / 55 —— 65	- 1650	29
	25 —— A —— 45 / 38 —— 60	- 380	44
26	30 —— A —— 50 / 38 —— 70	+ 4446	19
	20 —— A —— 45 / 55 —— 75	+ 1144	34
	25 —— A —— 45 / 38 —— 65	- 380	44
28	25 —— A —— 65 / 55 —— 70	- 1650	29
	20 —— A —— 55 / 50 —— 60	+ 465	43
32	25 —— 60 / 55 —— 40 — 30 —— 65	- 1650	29
	25 —— A —— 65 / 55 —— 80	- 1650	29
36	20 —— 65 / 55 —— 45 — 25 —— 40	- 1650	29
40	20 —— A —— 55 / 35 —— 60	+ 465	43
48	25 —— 60 / 55 —— 40 — 20 —— 65	- 1650	29
56	30 —— 60 / 55 —— 65 — 25 —— 70	- 1650	29
	20 —— A —— 55 / 25 —— 60	+ 465	43
60	25 —— 60 / 55 —— 50 — 20 —— 65	- 1650	29
72	38 —— 50 / 26 —— 60 — 25 —— 70	+ 4446	19

must be in virtually constant use, and although the writer made a very good living from his lathe, even that was not in use for an average of more than about 12 hours a week.

Moreover it is not possible to buy a modestly-priced lathe similar to that used by the writer whose lathe carries many self-made modifications for efficient and time-saving operation – such as a special clutch for high-speed threading (see Section 5), and a quick-acting self-ejecting rack operated tailstock, not to mention various other time-saving features which a majority of lathe makers steadfastly refuse to fit, and the fact that with all industrial lathes one would be forced to pay for a selective threading gearbox – which prevents the performance of some jobs unless gears are ordered.

Indeed, the average centre lathe is tolerated only because a majority of users remain blissfully unaware of its deficiencies – just as today some people are content with washing their clothes by wetting and banging with stones.

After much deliberation, a metric leadscrew of 3.5 mm pitch was considered to be the most advantageous for a lathe of about $3\frac{1}{2}$ in. centre height, and for the following reasons:
(1). 3.5 mm pitch approximates $7\frac{1}{4}$ threads/inch and is therefore slightly more robust than the 'standard' pitch of 3.0 mm. The half-nut threads are also somewhat stronger.
(2). The pitch of 3.5 mm slightly modifies 'pick-up' and this facilitates the automatic indexing of a range of multiple-start threads, as will be explained in Section 6
(3). The pitch of 3.5 mm used in conjunction with No. 17 (inverted) approximate translation ratio greatly simplifies quadrant gearing for threads/inch ratios. Indeed, as will be seen, in some instances

change to or from metric pitch ratios can be made by exchanging (or repositioning) only one gear.

THREADS/INCH FORMULA

Using our standard threads/inch formula for a leadscrew of 3.5 mm pitch in conjunction with No. 17 (inverted) approximation for 127/50, and substituting the known figures, we have:

$$\frac{\text{Drivers}}{\text{Driven}} = \frac{1}{\text{TPI}} \times \frac{10}{3.5} \times \frac{14}{5} \times \frac{39}{43}$$

(where TPI = threads/inch required) we find most figures cancel and can be rearranged to:

$$\frac{\text{Drivers}}{\text{Driven}} = \frac{8}{\text{TPI}} \times \frac{39}{43}$$

wherein the first fractional element (8/TPI) will be recognised as being the same as for a leadscrew of 8 tpi, and the second element (39/43) is the only requirement for translation from metric to English.

SOME EXAMPLES

The author frequently has call for threads of 8, 12 and 14 tpi when the 3.5 mm pitch leadscrew is in position, and the quadrant gearing for these threads is:

```
8 tpi    40— A —43
                39—40

12 tpi   40— A —43
                39—60

14 tpi   40— A —43
                39—70
```

also, halving the first 40 driver to 20 will give gearing for 16, 24 and 28 tpi. Moreover, gearing for 8 tpi is often preceded or followed by call for a thread of 4.0 or 3.5 mm pitch, and these pitches can be set thus:

```
4.0 mm pitch      40— A —43—35
                         39
```

3.5 mm pitch $\quad 40 - A - 43 - 40 \over 39$

where you will notice that only the last gear is altered or moved, and the 43 becomes a second idle gear, the 39 being conveniently left in position for changes back to threads/inch gearing. And, again, halving the 40 first driver to 20 gears for pitches of 2.0 mm and 1.75 mm.

BUILT-IN CONVERSION RATIOS

Although for reasons of lathe leadscrew pitch inaccuracies there is little point in seeking or in using metric/English conversion ratios of minimum error for quadrant gearing (unless the minimum error approximations happen to offer a more convenient approach, or simplify gearing), there is the point that when such ratios are to be built-in to a lathe as a permanent feature and are therefore outside the control of the operator, one may as well adopt the best ratio available and consistent with compactness. In this respect it so happens that some of the minimum error approximations are particularly suitable: notably Nos. 2, 3 and 7, (Table T5) wherein the average error is about $\frac{7}{8}$ in. a mile, or 14 millionths inch per inch.

An example of an in-line conversion ratio for changing the lead of a leadscrew

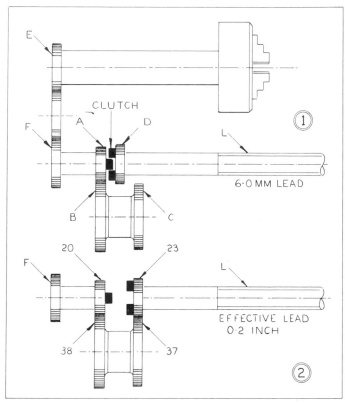

Fig. 11. An example of an in-line metric-English conversion ratio. At Fig. 11:1 the leadscrew of 6.0 mm. pitch is directly driven through the clutch, but at Fig. 11:2 the gearing shown gives the leadscrew an EFFECTIVE lead or pitch of 0.2 inch, i.e. 5 tpi to close limits. The error in the gearing is plus one part in 55,499 parts.

of 6.0 mm lead to an *effective* lead of 0.2 in., or 5 tpi, is given in the diagrams Fig. 11, 1 & 2 where, at Fig. 11.1 the drive to the leadscrew is straight through via the clutch, and at Fig. 11.2 the clutch has been disengaged and the 23 gear engaged with the 37 on the layshaft, with the result that the effective lead of the leadscrew is 0.2 in. to close limits when the 20T input gear is rotating at the same speed as that of the lathe spindle. This arrangement can of course be provided with a hand-lever for immediate selection between English and metric operation.

That this simple gearing should offer a metric/English conversion with an error of only plus one part in fifty-five thousand four hundred and 99 parts, or plus one thou. inch in $55\frac{1}{2}$ in. may seem surprising, but it is easily checked from our formula:

$$\text{Inch pitch} = \frac{\text{Drivers}}{\text{Driven}} \times \frac{5 \times \text{LSP}}{127}$$

with a 1-to-1 drive from the lathe spindle to gear A (Fig. 11.2) we have:

$$\frac{\text{Drivers}}{\text{Driven}} = \frac{20}{38} \times \frac{37}{23} \times \frac{30}{127} = \frac{11100}{55499}$$
$$= 0.2000035 \text{ inch}$$

showing an error of plus 3.5 millionths of an inch on pitch.

With the in-line translation ratio taking effect as in Fig. 11.2, gearing for threads/inch is determined by the formula as for a leadscrew of 5 tpi:

$$\frac{\text{Drivers}}{\text{Driven}} = \frac{\text{Effective leadscrew TPI}}{\text{TPI to be cut}}$$

In the example, Fig. 11:2 the effective leadscrew tpi is 5, accordingly, to set for a thread of 8 tpi, for example, would call for gearing in the ratio 5 to 8, say a 25 driver and a 40 driven, fitted to the input shaft at F.

In those instances where the lathe has a selective screw-cutting gearbox, the 25

wheel would of course be fitted to the gearbox 'output' shaft or spigot, and the box set as for a pitch of 6.0 mm, i.e. a 1 to 1 ratio. Thereafter, if the gearbox is set for a pitch of 3.0 mm the leadscrew would cut a thread of 16 tpi, and setting the box for a pitch of 2.0 mm would cut a thread of 24 tpi.

The table T9 details minimum error metric/English approximations for in-line gearing suitable for metric leadscrews from 7.0 mm to 3.0 mm pitch. In each example the lettering corresponds to that on Fig. 11. In those instances where teeth totals A plus B, and C plus D, differ by one or two teeth, it would of course be necessary to cut gears on slightly modified pitch diameters to suit the fixed centre distances.

In the opinion of the writer, however, and for the larger lathes, the provision of two leadscrews, one to English and one to metric standards, both permanently in position, would prove more straightforward than any amount of juggling with translation ratios. Moreover, pick-up problems (see Section 5) would be reduced to insignificance.

EXTRACTION OF 127/50 APPROXIMATIONS

As with the case of approximate gearing for a metric pitch from an English leadscrew, it is occasionally useful to extract or ascertain the approximation for 127/50 used in any approximate gear train for cutting threads/inch from a metric leadscrew. The approximation for 127/50 contained in any gearing may be found from the formula:

$$T = \frac{\text{Drivers}}{\text{Driven}} \times \frac{\text{TPI}}{1} \times \frac{\text{LSP}}{10}$$

where T = the approximation to be extracted, TPI = the nominal threads/inch

TABLE T9

IN-LINE METRIC-IMPERIAL CONVERSION GEARING FOR LEADSCREWS OF FROM 3·0 TO 7·0 MM PITCH OR LEAD

METRIC LEADSCREW PITCH – MM.	GEAR SIZES — READ IN CONJUNCTION WITH FIG. 1				EFFECTIVE NOMINAL TPI OF LEADSCREW GIVEN BY GEARING WITH 1:1 INPUT FROM LATHE SPINDLE	ACTUAL PITCH FROM GEARING ASSUMING A LEADSCREW OF PERFECT LEAD (INCH MEASURE)	LOW ERROR APPROXIMATIONS FOR THE RATIO 127:50 FROM WHICH GEARING IS DERIVED	EXACT ERROR IN THE APPROXIMATIONS
	INPUT A	DRIVEN B	DRIVER C	OUTPUT D				
7·0	32	31	29	33	4	0·249998	$\frac{16}{5} \times \frac{29}{31} \times \frac{28}{33}$	+ 1 IN 129921
6·0	20	38	37	23	5	0·2000035	$\frac{30}{19} \times \frac{37}{23}$	+ 1 IN 55499
6·0	31	28	26	34	5	0·1999933	$\frac{31}{17} \times \frac{39}{28}$	− I IN 30226
5·0	20	38	37	23	6	0·1666696	$\frac{30}{19} \times \frac{37}{23}$	+ 1 IN 55499
5·0	31	34	39	28	4	0·2499917	$\frac{31}{17} \times \frac{39}{28}$	− 1 IN 30226
4·0	29	31	28	33	8	0·124999	$\frac{16}{5} \times \frac{29}{31} \times \frac{28}{33}$	− 1 IN 129921
3·5	32	31	29	33	8	0·124999	$\frac{16}{5} \times \frac{29}{31} \times \frac{28}{33}$	− 1 IN 129921
3·0	20	38	37	23	10	0·1000017	$\frac{30}{19} \times \frac{37}{23}$	+ 1 IN 55499
3·0	31	28	26	34	10	0·0999966	$\frac{31}{17} \times \frac{39}{28}$	− 1 IN 30226

for which the gearing was devised, and LSP = the pitch of the leadscrew (in mm).

For example, gearing for a thread of 26 tpi from a leadscrew of 3.0 mm pitch may be:

$$30 - A - 50$$
$$38 - 70$$

what approximation (T) for 127/50 was used?

$$T = \frac{30}{50} \times \frac{38}{70} \times \frac{26}{1} \times \frac{3}{10} = \frac{39}{25} \times \frac{57}{35}$$

This is No. 19 (inverted) in table T5.

WORMS FOR GEARS SIZED BY MODULE

When it is necessary to cut a worm to mesh with a gear sized by module, and the leadscrew is of metric pitch, the quadrant gearing may be calculated from the general formula:

$$\frac{\text{Drivers}}{\text{Driven}} = \frac{22}{7} \times \frac{\text{Module}}{\text{Leadscrew pitch (mm)}}$$

For example, gearing for No. 1.25 module from a leadscrew of 3.0 mm pitch:

$$22/7 \times 1.25/3 = 11/7 \times 5/6$$

which, brought up to gear sizes could be set on a quadrant e.g.:

$$25 - A - 35$$
$$55 - 30$$

The metric pitch given = drivers/driven × leadscrew pitch = $25/35 \times 55/30 \times 3$ = 3.9285714 mm.

Circular pitch for No. 1.25 module = Pi × 1.25 = 3.9269907 mm.

The gearing therefore shows an error of plus 0.00158 mm, about 62 millionths inch. (The error arises from taking Pi as 22/7).

DIAMETRAL PITCH WORMS – METRIC LEADSCREWS

A theoretically accurate formula for gearing a metric leadscrew to cut worms sized by diametral pitch reads:

$$\frac{\text{Drivers}}{\text{Driven}}$$

$$= \frac{10}{\text{LSP (mm)}} \times \frac{3.1415926}{DP} \times \frac{127}{50}$$

where DP = the diametral pitch number and LSP = the leadscrew pitch in mm.

Here of course the seven-digit decimal portion for Pi makes it quite impossible to modify to change gear sizes. However it so happens that 3.1415926 (Pi) × 127 = 398.98226, which rounds up to 399 with an error of plus 1 in 22491. Moreover 399 factorises into 3 × 7 × 19, and using this approximation, the gearing formula (after cancellations where possible) can be rearranged to

$$\frac{\text{Drivers}}{\text{Driven}} = \frac{3}{\text{LSP (mm)}} \times \frac{19}{DP} \times \frac{7}{5}$$

Using this formula for a leadscrew of 3.0 mm pitch, the first of the three elements cancels out, and, for example, gearing for a No.20 DP worm becomes:

$$\frac{\text{Drivers}}{\text{Driven}} = \frac{19}{20} \times \frac{7}{5}$$

which when multiplied up to available change gear sizes would set on a quadrant for example:

$$38 - A - 50$$
$$70 - 40$$

This gearing, under perfect conditions, would cut a No. 20 DP worm with a pitch error of plus 7 millionths inch.

A less accurate formula which may nevertheless sometimes prove useful reads:

$$\frac{\text{Drivers}}{\text{Driven}} = \frac{4}{\text{LSP (mm)}} \times \frac{20}{DP}$$

with which, for example, gearing for No. 20 DP resolves to a simple 40/30 with a leadscrew of 3.0 mm pitch. With this gearing, under perfect conditions, the pitch error for a No. 20 DP worm is plus 4/10 thou. inch.

CHANGE GEAR CALCULATIONS BY APPROXIMATION

Although it is not possible to arrange *accurate* gearing for any particular thread unless a gear is available having a number of teeth which is the same as, or an exact multiple of, the tpi or metric pitch number required, a reasonable compromise can often be found by approximation.

For example, a thread of 19 tpi ordinarily requires a 38 gear special to a set rising by fives, and with a leadscrew of 8 tpi the gearing is 20/38 × 40/50, but if a 38 gear is not available we have to see what can be done with the gears by fives only. One way of finding an approximation ratio is as follows:

First find what we may term the 'basic ratio', in this example 8/19 which equals a step-down ratio of 1 to 2.375. Now, by repeated addition of 2.375 with a calculator, (which for example with a Casio LC 826 can be done at the rate of 80 a minute with the time to examine each new set of figures) continue until a figure is reached in which the decimal portion is small enough to be ignored, or is large enough to be counted as 1 (one) for adding to the integral figures. In this case, at the 21st addition, we get a figure of 49.875. Rounding this up to 50, we have a driver of 21 (the 'count' figure when adding) and a driven wheel of 50, and 21/50, after factorising and multiplying up to available change gear sizes, offers quadrant gearing:

$$35 - A - 50$$
$$45 - 75$$

which gives a theoretical pitch error of minus 0.000132 in., or, roundly, $2\frac{1}{2}$ thou. in. per inch of thread.

If the repeated addition is continued to the 27th we obtain a reading 64.125, and, ignoring the decimal portion, we have 27/64 which factorises into 3/8 × 9/8, offering change gears that would set on a lathe:

$$25 - 40 \quad 30 - 50$$
$$45 - 40$$

This is a more accurate ratio, showing a pitch error of plus 0.000101 in., but it could not be set without two 40 gears.

Similarly, the basic ratio for a thread of $11\frac{1}{2}$ tpi from a leadscrew of 8 tpi is 16/23 = 1 to 1.4375, and by repeated addition we have, after the 24th.:

25th. addition 25/35.9375 = 25/36
34th. addition 34/48.875 = 34/49
39th. addition 39/56.0625 = 39/56
57th. addition 57/81.9375 = 57/82

The 25th and 39th additions hold primes within our range, and since the higher the multiplier the less the error, we will try the 39th: 39/56 which factorises to 3/7 × 13/8 and when brought to gear sizes offers a quadrant setting:

$$30 - A - 40$$
$$65 - 70$$

With this gearing there is a theoretical pitch error of minus 0.000097 inch.

ENGLISH – METRIC APPROXIMATIONS

The following basic ratio formulas for gearing an English leadscrew for a metric pitch by approximation are arranged for keying on a calculator.

Leadscrew Threads/inch	Key*
4	127/(20 × P)
5	127/(25 × P)
6	127/(30 × P)
8	127/(40 × P)
10	127/(50 × P)

*On a Casio LC 826 one would key for example for 2.2 mm pitch from a leadscrew of 8tpi (40 × 2.2) ÷ 127

In each example P represents the metric pitch for which approximate gearing is to be found.

For example, required a basic ratio for a thread of 2.2 mm pitch from a leadscrew of 8 threads/inch.:
127/(40 × 2.2) = 1.4431818 (store in memory).
At the ninth addition we have 12.9888634, rounding to 13 offers gearing in the ratio driver/driven = 9/13 which brought to gear sizes = 45/65. The pitch given = 2.1980769 mm, a minus error of 0.0019231 mm, or about 75 millionths/inch.

METRIC – ENGLISH APPROXIMATIONS

The following basic ratio formulas for gearing a metric leadscrew for threads/inch by approximation are arranged for keying on a calculator.

Leadscrew Pitch Millimetres	Key
3.0	(15 × TPI)/127
3.5	(35 × TPI)/254
4.0	(20 × TPI)/127
5.0	(25 × TPI)/127
6.0	(30 × TPI)/127
7.0	(35 × TPI)/127

For example, required a basic ratio for a thread of 25 tpi from a leadscrew of 3.0 mm pitch:

$$(15 \times 25)/127 = 2.9527559$$

At addition 22 we have 64.960629, rounding this up to 65 therefore gives a basic ratio of 22/65 = 2/5 × 11/13, say:

```
.20 — A — 50
     55 — 65
```

This gearing gives a pitch of 0.0399757 inch compared with a true pitch of 0.04 inch, the error being minus 24.3 millionths of an inch, or about 6/10 thou. in. per inch, assuming an error-free leadscrew.

Occasionally when resolving a gearing formula we may encounter values for which gears are not available, such as for example 74/75 (2/5 × 37/15) which would call for a 37 or 74 gear outside our range. However, if 1 (one) is added to both numerator and denominator of 74/75 we have 75/76 = 5/2 × 15/38 which would fall within the Myford range of gears by fives plus one 38. The error introduced by the addition of 1 (one) to numerator and denominator is plus one part in 5624 parts.

An interesting example of the effect of modifying by addition occurs with No. 23 approximation for the English/metric conversion ratio 50/127 (Section 3): 22/49 × 50/57 which has an error of plus 1 in 2793. But if 1 (one) is added to numerator 50 and denominator 49, we have 22/50 × 51/57 which resolves to No.10 approximation: 11/25 × 17/19, having an error of minus 1 in 23750, an error $8\frac{1}{2}$ times less than that given by No. 23 approximation.

Problems and Analysis of Repeat Pick-up

(Avoidance of 'crossed threading')

A sound knowledge of the relationship between the leadscrew threads, leadscrew gearing, and the helices on a component being threaded can result in the saving of a very considerable amount of time, especially when engaged on quantity production of threads of any significant length, where there is a potential saving, not of the odd half-hour, but of whole working weeks. As has been explained, threads are lathe screwcut by taking a series of passes of progressively increasing depth. What we now have to consider is the fact that unless certain conditions are met, a threading tool will not always follow the first helix trace on subsequent passes except in a limited number of instances.

As already hinted, *repeat pick-up* or 'pick-up' is the term commonly used to indicate that a threading tool is repeatedly and exactly following the groove of a thread being cut. For our purposes here it will occasionally be useful to borrow an expression from the vocabulary of the electrical world, and to refer to repeat pick-up as *synchronous* working, or to state that an exactly following threading tool is in *synchronism* with the thread being cut.

(For lathes of specialised design provided with a dog-clutch arrange-

ment giving high-speed repeat pick-up for *all* pitches without special intervention by the operator, see page 83).

Let us first deal with those few cases where, with ordinary lathes as distinct from those with dog-clutch control, repeat pick-up is assured.

If 8 tpi is being cut with a leadscrew of 8 tpi, and the first cutting pass has been made, then for the next, and all subsequent cutting passes, the half-nuts may be re-engaged at any moment when their threads happen to coincide with the leadscrew threads, and the tool will follow the original helix. This may be demonstrated after taking the first pass, by stopping the lathe, taking up gearing backlash, and making trial half-nut engagements with the tool just clear of the workpiece and positioned at various intervals along the length of the helix: at each re-engagement of the half-nuts, the tool point will be seen to exactly coincide with the helix. In fact, with the provision that the leadscrew is not de-geared and the tumbler-reverse has not been interfered with, you will find it impossible to re-engage the half-nuts in an adverse position. The same conditions will hold if the workpiece (and lathe spindle) is turned through part of a revolution, or is slowly revolving whilst trial half-nut engagements are being made.

71

Similar trial half-nut engagements with a workpiece having 16, 24, 32 or 40 tpi and a leadscrew of 8 tpi will also show the impossibility of engaging the half-nuts in an adverse position. 16 tpi picks up at 2 component thread intervals, 24 tpi at 3 component thread intervals, and so on; consequently we may say that if the number of threads/inch being cut is the same as, or an exact multiple of the number of threads per inch of the lead-screw, no special precautions are required to ensure pick-up.

Now let us suppose that with a lead-screw of 8 tpi, a 9 tpi helix has just been traced upon a piece of stock of just over one inch in length (gear ratio 40 to 45), but the lathe has been stopped at the termination of the trace before disengaging the half-nuts. The tool point will of course coincide with the trace at its termination. But if the tool is now withdrawn just clear of the workpiece, and the half nuts disengaged, subsequent trial re-engagements along the length of the workpiece will show a series of adverse tool positions until the carriage has been traversed exactly one inch to the right (8 leadscrew threads) at which position the tool point will again coincide with the ninth thread groove. It will also be seen that at a distance of half an inch from the terminal position of the trace, the tool point will be exactly mid-way between two thread traces. If the experiment is made, and the lathe set in very slow motion, it will be easy to see that the chances of re-engaging the half-nuts at a correct pick-up moment are remote. The reason for these adverse tool positions in relation to the helix lies in the fact that the number of work threads to each single leadscrew thread is fractional until the eighth leadscrew thread is reached. One leadscrew thread 'holds' $\frac{9}{8} = 1\frac{1}{8}$ component thread turns, two leadscrew threads

hold $\frac{9}{8} \times 2 = 2\frac{1}{4}$ component thread turns, and so on up to seven leadscrew threads holding $\frac{9}{8} \times 7 = 7\frac{7}{8}$ component thread turns, and $\frac{9}{8} \times 8 = 9$ complete component thread turns in one inch, or 8 leadscrew threads.

However, from the foregoing it will not be difficult to perceive that synchronism or pick-up could always be assured (for cutting odd number tpi from standard English leadscrews) by the following sequences:

(1). Stop the lathe on completion of each cutting pass, and with the half-nuts engaged.

(2). When the lathe has stopped, disengage the half-nuts. Run the carriage to the right through a sufficient number of whole inches (groups of 8 leadscrew threads) for the threading tool to clear the start of the thread. Re-engage the half-nuts.

(3). Adjust tool depthing. Set the lathe in motion for the next cutting pass.

NOTE: Right carriage repositioning traverse measurement as at (2) should always be made from where the carriage last ceased screwcutting traverse, and all cutting passes must be made in full.

For want of a better term, this will be referred to as the 'Stop and Reset' method or as 'Rack & Pinion Resetting' (see also page 89).

We may therefore state that when a thread to be cut is not an exact multiple of the leadscrew threads, correct pick-up occurs only at those positions on a lead-screw where a complete whole number of threads to be cut corresponds to a complete whole number of leadscrew threads, English or metric. For future reference, these measurements or distances will be referred to as the 'synchronous distance' (SD) or as 'minimum pick-up'.

QUESTION:

Is it always possible to stop the lathe spindle and leadscrew on completion of each cutting pass, and to run the carriage and tool clear of the workpiece, then to re-engage the half-nuts at a synchronous distance to hold pick-up for the next cutting pass?

1. There will be an SD for every conceivable combination of gearing between a component thread and leadscrew of either language, but sometimes, in relationship to the length of the thread being cut, the SD will be of too great a length to render it of any practical use. Occasionally the SD will exceed the entire leadscrew length, then of course, the SD cannot be used at all.

2. When common threads are being cut in the same language as that of the leadscrew (English with English, or metric with metric) and the correct gearing (as distinct from approximation gearing) is being used, SD's are more often enough to be of practical use.

3. When a thread being cut is in a language opposite to that of the leadscrew (metric with English or vice versa) the SD's are sometimes short enough to be of practical use for carriage and tool repositioning. Generally speaking, when a thread being cut is in a language opposite to that of the leadscrew and approximate gearing is being used for the ratio 50/127 or for 127/50, those approximations of low value (maximum error) offer the shortest SD's, and those of high value (minimum error) offer the longest SD's.

HOW IS THE SD CALCULATED?

With either an English or metric leadscrew of **ANY** pitch, the SD, or minimum pick-up for **ANY** component thread pitch (English, metric, DP or module etc.) can be ascertained by reducing the number of teeth in the driving and driven gears to their lowest integral, or whole number terms. The driving figure (numerator) will then give the minimum whole number of leadscrew threads (LST), and the driven figure (denominator) the minimum proportionate whole number of work thread turns (WT). The synchronous distance (SD), or minimum pick up, is then found by multiplying the leadscrew pitch (expressed by inch or metric measure) by the minimum LST figure. Hence we may say, with drivers and driven reduced to lowest terms:

$$\frac{\text{Drivers}}{\text{Driven}} = \frac{\text{LST}}{\text{WT}}$$

and LST × leadscrew pitch = SD

It is important to note that this formula will give minimum pick-up or SD for **ALL** ratios, including English/metric conversion gearing (and vice versa) and any other approximate gearing.

Some examples will now be given.

Example 1. Leadscrew 8 tpi, geared to cut 19 tpi

$$20 - A - 38$$
$$40 - 50$$

$$\frac{\text{Drivers}}{\text{Driven}} = \frac{20}{50} \times \frac{40}{38} = \frac{8 \text{ LST}}{19 \text{ WT}}$$

and 8 LST of $\frac{1}{8}$ in. pitch = 1 inch SD.

Example 2. Leadscrew 8 tpi, geared to cut approximately 19 tpi (38 gear not available):

$$35 - A - 50$$
$$45 - 75$$

$$\frac{\text{Drivers}}{\text{Driven}} = \frac{35}{50} \times \frac{45}{75} = \frac{21 \text{ LST}}{50 \text{ WT}}$$

and 21 LST of $\frac{1}{8}$ in. pitch = $2\frac{5}{8}$ inch SD.

Example 3. Leadscrew 10 tpi, geared to cut 13 tpi:

$$40 - A - 20$$
$$25 - 65$$

$$\frac{\text{Drivers}}{\text{Driven}} = \frac{40}{20} \times \frac{26}{65} = \frac{10 \text{ LST}}{13 \text{ WT}}$$

and 10 LST of $\frac{1}{10}$ in. *pitch* = 1 inch SD.

Example 4. Leadscrew 8 tpi, geared to cut a thread of 2.0 mm pitch by approximate gearing (using No. 11 approximation ratio for 50/127)

$$26 - \text{A} - 38$$
$$45 - 40$$

$$\frac{\text{Drivers}}{\text{Driven}} = \frac{26}{38} \times \frac{45}{40} = \frac{117 \text{ LST}}{152 \text{ WT}}$$

and 117 LST of $\frac{1}{8}$ in. pitch = $14\frac{5}{8}$ in. SD.

Example 5. Leadscrew 8 tpi, geared to cut 1.75 mm pitch with 50/127 translation ratio:

$$40 - \text{A} - 20$$
$$35 - 127$$

$$\frac{\text{Drivers}}{\text{Driven}} = \frac{40}{20} \times \frac{35}{127} = \frac{70 \text{ LST}}{127 \text{ WT}}$$

and 70 LST of $\frac{1}{8}$ in. pitch = $8\frac{3}{4}$ in. SD.

Example 6. Leadscrew 8 tpi, geared to cut 3.0 mm pitch by approximate gearing (using No. 3 approximation for 50/127):

$$23 - \text{A} - 25$$
$$38 - 37$$

$$\frac{\text{Drivers}}{\text{Driven}} = \frac{23}{25} \times \frac{38}{37} = \frac{874 \text{ LST}}{925 \text{ WT}}$$

and 874 LST of $\frac{1}{8}$ in. pitch = $109\frac{1}{4}$ in. SD.

In this example it may be thought that with a minimum SD of about nine feet, something must be wrong, but since none of the gears cancel or are further reducable, the LST figure is correct, and it requires exactly 925 **NOMINAL** 3.0 mm pitch turns to span the same distance of $109\frac{1}{4}$ inches. (The metric pitch given by the gearing is 2.9999457 mm).

Example 7. Leadscrew 3.0 mm pitch, geared to cut a thread of 0.75 mm pitch:

$$20 - \text{A} - 40$$
$$25 - 50$$

$$\frac{\text{Drivers}}{\text{Driven}} = \frac{20}{40} \times \frac{25}{50} = \frac{1 \text{ LST}}{4 \text{ WT}}$$

and 1 LST of 3.0 mm pitch = 3.0 mm SD.

Example 8. Leadscrew 3.0 mm pitch, geared to cut a thread of 12 tpi using No. 19 approximation for 127/50:

$$38 - \text{A} - 50$$
$$65 - 70$$

$$\frac{\text{Drivers}}{\text{Driven}} = \frac{38}{50} \times \frac{65}{70} = \frac{247 \text{ LST}}{350 \text{ WT}}$$

and 247 LST of 3.0 mm pitch = 741 mm (About 29 inches SD).

Example 9. A special leadscrew of $7\frac{1}{3}$ tpi is used (for special reasons) to cut a thread of $5\frac{1}{2}$ tpi, by quadrant gearing:

$$40 - \text{A} - \text{A} - 30$$

What is minimum pick-up or SD?

$$\frac{\text{Drivers}}{\text{Driven}} = \frac{40}{30} = \frac{4 \text{ LST}}{3 \text{ WT}}$$

and 4 leadscrew threads of $7\frac{1}{3}$ tpi (0.1363636 inch pitch) = 0.5454 inch SD.

In examples of this type (9) (which the author has actually used) a precise SD measurement is unnecessary because, as will be seen, the sole purpose of ascertaining the SD is to find the position (on a 'dead' leadscrew) at which to re-engage the half-nuts to hold pick-up for a thread being cut, and one is hardly likely to drop in at the third or fifth leadscrew thread with the knowledge that re-engagement must be made (in this example) at the fourth (or at the eighth, twelfth, or sixteenth, according to length of thread).

EFFECT OF LEADSCREW ERRORS

At this juncture it may be asked if errors in pitch of a leadscrew can affect pick-up. The answer is no. Pick-up depends *only* upon the *number of leadscrew thread turns* (English or metric) and if a leadscrew is 'slow' (too fine) or 'fast' (too coarse) then the pitch of a screw being cut will be correspondingly too fine or too coarse, but the proportionate work-thread turns to leadscrew-thread-turns will exactly hold good.

NOTE: In the event of English threads being cut from a metric leadscrew (of any pitch) and with the exact 127/50 translation ratio, as 127 is a prime, the *minimum* LST figure can never be less than 127. Hence if a leadscrew is of 3.0 mm pitch, the SD will be 381 mm (15 in.), and for a leadscrew of 6.0 mm pitch, SD will of course be 762 mm (30 in.).

AN ARITHMETICAL PARADOX

When calculating threads/inch quadrant gearing with an English leadscrew, the leadscrew threads/inch appears as the numerator in the formula:

$$\frac{\text{Drivers}}{\text{Driven}} = \frac{\text{Threads/inch of leadscrew}}{\text{Threads/inch to be cut}}$$

and in any resulting basic ratio, such as $\frac{8}{9}$ for 9 tpi from a leadscrew of 8 tpi, the numerator 8 remains as the leadscrew-thread-TURNS-to-component-thread-TURNS RATIO figure for calculating pick-up or SD's.

On the other hand when a metric leadscrew and thread to be cut are both expressed in terms of PITCH, we have the formula:

$$\frac{\text{Drivers}}{\text{Driven}} = \frac{\text{Metric pitch to be cut}}{\text{Metric pitch of leadscrew}}$$

and, for example, we use a basic driver 5 and driven 6 for cutting a pitch of 5.0 mm from a leadscrew of 6.0 mm pitch, yet when the basic ratio $\frac{5}{6}$ is used for assessing pick-up, the pitch-to-be-cut figure (5) becomes the LST numerator figure, and the leadscrew pitch in mm (6) becomes the component-thread-turns denominator figure:

$$\frac{5 \text{ LST of } 6.0 \text{ mm pitch span } 5 \times 6}{6 \text{ WT of } 5.0 \text{ mm pitch span } 6 \times 5}$$

$$\frac{= 30 \text{ mm}}{= 30 \text{ mm}} \text{ SD}$$

Much confusion can be avoided by noting and accepting this paradoxical situation, which appears to defy explanation other than by example.

We may note, however, that the same interchange or inversion occurs when an English leadscrew and the thread to be cut are both expressed in inch pitch for which the basic gear ratio is derived from the formula:

$$\frac{\text{Drivers}}{\text{Driven}} = \frac{\text{Pitch by inch measure to be cut}}{\text{Pitch of LS by inch measure}}$$

whereby, for example, gearing for a thread of 10 tpi (0.1 in pitch) from a leadscrew of 8 tpi (0.125 in pitch) resolves to 0.1 driver/0.125 driven, which, raised to least whole numbers shows a 4 driver and a 5 driven. Yet for determining pick-up or SD we have:

$$\frac{4 \text{ LST} \times 0.125 \text{ in.} = 0.5 \text{ in.}}{5 \text{ WT} \times 0.100 \text{ in.} = 0.5 \text{ in.}} \text{ SD}$$

wherein, of course, what was the numerator pitch-to-be-cut figure has to be used as the LST figure.

REPEAT PICK-UP METHODS. 'COMPLETE REVERSAL'

Frequently with the type of lathe more generally used today, when the pitch of a thread to be cut shows an awkward relationship to the pitch of the leadscrew, screwcutting is carried out with the half-nuts remaining in engagement until completion of the thread, both lathe spindle and leadscrew being reversed to reposition the carriage and tool for every fresh cutting pass. For this method the lathe requires a reverse drive for the headstock, and (as with the stop & reset method) extra care is required to stop in time to prevent the tool from running into a shoulder. When the lathe is reversed with the tool clear of the component, the leadscrew drives the carriage back to its starting position, and correct pick-up is held because each return pass is a pure reversal of each cutting pass. For future reference this procedure will be referred to as THE COMPLETE REVERSAL METHOD.

It may be as well to mention here that when using the complete reversal method on the more common type of lathe, the lathe spindle must be reversed to drive the screwcutting gear train backwards. This is emphasised because some text books can give the impression that the tumbler-reverse can be used to return the carriage by reversing only the leadscrew with the half-nuts engaged. For example, "Workshop Technology" (Part 1 Fifth Ed. 1971) by Dr. W.A.J. Chapman states, at pages 359-360 "Some lathes have a lever on the apron which reverses the leadscrew without reversing the machine, and this method is about the most convenient of all for bringing the tool back to its starting position." What Dr. Chapman is referring to here (although no explanation is given) is a special reversible single-tooth dog-clutch positioned at the input end of screwcutting gear trains, and not to some special arrangement for operating the tumbler-reverse from the apron, as the uninstructed may easily be led to believe. I also recall reading somewhere that, quote: "A tumbler-reverse is handy for repositioning the carriage when cutting long threads." This is quite wrong. If a tumbler-reverse is de-geared at any time during the screwcutting, then correct pick-up will be lost. Not only that, far from being 'handy', waiting for a reversed leadscrew to reposition a lathe carriage over any significant distance is a waste of time, and best avoided whenever possible. The

special dog-clutch here referred to is described under DOG-CLUTCH CONTROL at page 83.

TRADITIONAL PICK-UP AIDS: THE LEADSCREW INDICATOR

We have seen that when threading, for example, 7, 9, 11, 13, 19 tpi with a standard English leadscrew, the lathe may be stopped at the end of each cutting pass, the half-nuts disengaged, and pick-up regained by returning the carriage to the right through a sufficient number of whole inches for the tool to clear the component, then re-engaging the half-nuts for the next cutting pass. Similarly, of course, for even number tpi such as 10, 14, 18, 22, 26, the carriage requires right traverse through steps of half an inch, and threads divisible by 4 such as 12 and 20 will pick up at $\frac{1}{4}$ inch intervals. The disadvantage of this method lies in the lathe having to be stopped before disengaging the half-nuts, and this led to the development of the THREAD DIAL INDICATOR, also known as a 'THREAD CHASING DIAL', or, perhaps best, as a 'LEADSCREW INDICATOR' (LSI) which obviates the trouble of having to stop the lathe at the termination of each cutting pass. As will be seen, the LSI registers favourable half-nut engagement moments with the lathe running and, in effect, a LSI integrates leadscrew revolutions with carriage movement in such a way that favourable half-nut engaging moments are repeatedly registered from any convenient carriage and tool starting position.

An indicator is shown diagrammatically in Fig. 12. The arrangement consists essentially of a worm wheel engaging with the leadscrew, the worm wheel spindle being vertically mounted and fitted at its upper end with a dial having four main divisions numbered 1-2-3-4, and four un-numbered sub-divisions. The spindle is free to revolve in a bearer mounted on the lathe carriage.

The number of teeth on the worm wheel is always some exact multiple of the number of threads on the leadscrew designated either by threads to the inch or by metric pitch. In this example a wheel of 16 teeth is shown engaging with a lead-screw of 8 threads to the inch, such as would be fitted to a small lathe. Larger lathes with leadscrews of 4 threads/inch may have a similar 16 teeth worm wheel, but there would be slight differences in interpreting the indicator readings, as will be seen.

During the initial fitting of a leadscrew indicator, lateral fixing adjustments are made so that when the half-nuts are engaged, any one of the indicator gradua-tions falls exactly opposite a fiducial mark, indicated at *P* on the diagram.

ACTION OF THE INDICATOR

If, when the leadscrew is stationary, the carriage is moved along the lathe bed, the indicator worm wheel treats the lead-screw as a rack, and the indicator dial will revolve. As the leadscrew in our diagram is presumed to be of 8 tpi and the worm wheel has 16 teeth, each exact turn of the indicator dial will measure a carriage traverse movement of 2 in., half a revolu-tion therefore registers a movement of one in., and so on, down to the one-eighth divisions which show a movement of two leadscrew threads, or $\frac{1}{4}$ in.

On the other hand, if the carriage is stationary, and the leadscrew is rotating, the leadscrew will act as a worm, and will drive the indicator, whereupon each complete turn of the indicator will show that the leadscrew has made 16 whole turns, half a turn of the indicator, 8 turns of the leadscrew, and so on.

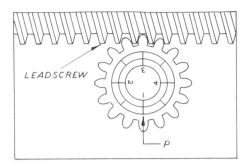

Fig. 12. A leadscrew indicator for an English leadscrew.

Now suppose a $1\frac{3}{4}$ in. length of 9 tpi is being cut, and the total carriage movement is 2 in., giving a $\frac{1}{4}$ in. tool starting clearance. For explanatory purposes, let the first cutting pass be taken when the LSI reads 1, (one) stop the lathe on completion of the pass, and disengage the half-nuts. If the carriage is now returned through a sufficient distance to the right to cause the LSI to revolve from 1 to 1, showing a 2 in. traverse, the half-nuts can be re-engaged ready for the next threading pass, and pick-up would be assured.

At this juncture we may note that when the half-nuts are engaged and the carriage is making a threading pass, the LSI dial remains stationary because that portion of the worm wheel which is engaging with the leadscrew merely acts in the same way as would a few threads of a nut similarly engaged and moving with the carriage.

Let us now see what happens to the LSI when cutting the same $1\frac{3}{4}$ in. length of 9 tpi thread without stopping the lathe on completion of each threading pass.

As before, with the carriage in its starting position, we wait until the lead-screw has turned the indicator to reading 1, then engage the half-nuts for the threading pass. On disengaging the half-nuts at the end of the threading pass, the indicator dial commences to revolve, and during the time it takes to reposition the carriage 2 inches to the right, the leadscrew may make two revolutions, so the indicator will, at the instant of ceasing the right traverse, read 1 again, showing one whole turn plus one eighth of a turn for the two leadscrew revolutions. Favourable half-nut engagement positions or moments occur only at each 8 revolutions of the leadscrew, or at one inch or multiples of one inch of carriage travel, so, assuming the carriage cannot be moved a further $\frac{3}{4}$ inch to the right, the lathe is left running until the leadscrew has made 6 more turns which will be shown by the indicator moving through a further $\frac{3}{8}$ turn to reading 3, and the half-nuts may be re-engaged. We may notice also that from the time of disengaging the half-nuts to the time of re-engaging the work will have made 9 revolutions and the leadscrew 8.

If for any reason the half nuts are not engaged on the first favourable reading 3, then, of course, the leadscrew may be allowed to revolve until the indicator again reads 1, showing 8 more leadscrew revolutions.

An indicator such as that just described which makes one dial revolution for each 2 inches of travel could be termed a "2-inch indicator".

When the leadscrew is of four tpi and the indicator has the same number of teeth – 16 – the dial will show four inches of carriage travel per revolution, and we have a "four-inch indicator," although the number of leadscrew threads for each revolution of the dial will remain at 16. In case of doubt, of course, the amount of carriage travel to each revolution of an indicator can be found with a rule.

For a two-inch indicator, when cutting odd threads such as 7, 9, 11 and so on, it is well to form the habit of engaging the half-nuts only at the odd dial numbers 1 or 3, (one inch of travel). For even numbers divisible by 2, the half-nuts may be engaged on any numbered division, 1, 2, 3 or 4. For even numbers divisible by 4, such as 12 tpi with three whole threads and 20 tpi with five whole threads to each *two* leadscrew threads, the half-nuts may be engaged at any $\frac{1}{8}$ dial division.

With a four-inch indicator and a 4 tpi leadscrew, odd threads may be picked up at any numbered dial division, 1, 2, 3 or 4, showing inches of carriage travel. Threads divisible by 2, such as 6, 10, 14 will pick up at any $\frac{1}{8}$ revolution of the dial, showing two leadscrew threads, or half inches of travel. For threads divisible by 4, such as 4, 8, 12, 16, 20, and upwards the use of the indicator is not strictly necessary, although it is sometimes handy to watch an indicator to avoid attempts at engaging the half-nuts on to the leadscrew thread crests: an event which can sometimes result in spoiled work by giving a false start to a threading pass. Half-nut thread crests can grip leadscrew thread crests with sufficient force to traverse a lathe carriage.

The limit of usefulness of leadscrew indicators is reached when the SD of the thread to be cut is equal to one revolution of the indicator dial, for example $11\frac{1}{2}$ tpi and similar 'half numbers' for a 2-inch indicator, and $11\frac{1}{4}$ and similar 'quarter numbers' for a 4-inch indicator. $11\frac{1}{2}$ tpi shows 23 whole threads in 2 inches, and $11\frac{1}{4}$ tpi = 45 whole threads in 4 inches.

With a majority of present lathes customary method for holding pick-up when the SD exceeds the capacity of a leadscrew indicator is to work by the complete reversal routine. It is also customary (although not always necessary, q.v.) to work by this method when cutting metric threads with an English leadscrew and vice versa.

Worm threads sized by DP or module also have SD's outside the scope of leadscrew indicators. For example the actual number of threads/inch held by a worm to mesh with a No. 20 DP gear is $6\frac{4}{11}$:35 whole threads minimum in $5\frac{1}{2}$ inches.

NOTE: Leadscrew indicators are of no use unless the gearing is exact (i.e. is not an approximation) for the pitch being cut. If the SD exceeds the indicator travel, then the leadscrew indicator cannot be used.

For example, if, for lack of a 38 gear, a nominal 19 tpi is being cut with a leadscrew of 8 tpi with gearing:

$$35 - A - 50$$
$$45 - 75$$

minimum pick-up, or SD, is, as we have seen, $2\frac{5}{8}$ inches, and is therefore outside the scope of a leadscrew indicator that will register only in a maximum of 2-inch groups, or in exact sub-multiples of 2 inches.

GEARED LEADSCREW INDICATORS

The pending metrication led to the development of geared leadscrew indicators, their design probably being based upon the fact that when cutting metric threads from an English leadscrew used in conjunction with the exact 50/127 translation ratio, SD, or minimum pick-up in INCHES is equal to 5 multiplied by the pitch of the thread to be cut expressed in millimetres.

For example, with a leadscrew of 4 tpi and the 50/127 translation ratio, a pitch of 1.0 mm picks up at 5 inch intervals, or at each 20 revolutions of the leadscrew, 2.0 mm pitch at 10 in or 40 leadscrew revs and so on.

Hence a 4-inch leadscrew indicator having an additional dial driven through reduction gearing in the ratio 1 to 5 would register 20 in. (80 LS revs) in one turn of the dial, 10 in. (40 LS revs) in half a turn, 5 in. (20 LS revs) in $\frac{1}{4}$ turn, $2\frac{1}{2}$ in. (10 LS revs) in $\frac{1}{8}$ turn, and $1\frac{1}{4}$ in. (5 LS revs) in $\frac{1}{16}$ turn, thus making the geared dial suitable for registering pick-up for 4.0, 2.0, 1.0, 0.5 and 0.25 mm pitches. A miniature selective gearbox is used to change the geared LSI ratio for registering other pitches. For example a reduction in the ratio 4 to 25 is required for a dial to register pick-up for 5.0, 2.5 and 1.25 mm. pitches at 100, 50 and 25 leadscrew revolutions: 25 in,. $12\frac{1}{2}$ in., and $6\frac{1}{4}$ in. respectively.

However, waiting for a geared LSI dial to show pick-up for the coarser metric pitches could waste considerable time: a leadscrew rotating at say 100 rpm could waste up to a whole minute between each threading pass when cutting a 5.0 mm pitch, which picks up only at every 100 leadscrew revolutions, although waiting time will decrease with the length of the thread to be cut, because during the non-cutting return passes when rack traversing to the right, the leadscrew indicator dial will rotate one turn for every 25 inches of right traverse, so if a thread is of 22 in. length (88 LST), then after repositioning at the right, the dial will have only to count the remaining leadscrew thread/revs, less the few revolutions the leadscrew will have made during the right traversing. For example, 22 in. = 88 LST, add 5 for the revs made by the leadscrew during right traverse = 93, and 100-93 = 7, i.e. 7 leadscrew revolutions to be made before the LSI dial registers correct pick-up, for which at 100 rpm the operator would have about 11 seconds in hand. But of course, if the first favourable moment is missed, then the total waiting time would be 71 seconds.

To minimise this potential waste of time, these geared leadscrew indicators have been used in conjunction with automatically disengaging half-nuts. Threading passes are commenced at say (geared) LSI dial reading 1. On completion of the threading pass, the half-nuts automatically disengage. The lathe is stopped, the carriage traversed to the right, and the lathe run in reverse until the LSI dial has been driven back to its original starting reading, whereupon, of course, the half-nuts are engaged and the lathe run in forward motion for the next cutting pass: all of which takes about the same time as the complete reversal method, but with the advantage that fear of over-run is eliminated.

However, geared leadscrew indicators (and indeed ordinary LSI's) probably will not long survive in competition with much more efficient methods for holding pick-up, so a deeper analysis of their characteristics seems unnecessary.

SPECIAL APPLICATION OF LEADSCREW INDICATOR

When threading up to shoulders or to the base of a blind bore in the conventional way with the customary type of lathe and (ungeared) leadscrew indicator, there is an ever present risk of the carriage over-running, breaking the tool and spoiling the work. To avoid this risk, and to allow of somewhat faster working with less anxiety, some turners are in the habit of cutting such threads by running the lathe in reverse, using an inverted threading tool, and taking the cutting passes from left to right. The non-cutting return passes are then made by hand traversing from right to left against a dead stop, and the leadscrew indicator used to show correct pick-up. For this kind of working the lathe is run in reverse throughout, and the thread being cut must be of a kind for

which a leadscrew indicator is suitable for use in the normal way.

SHORT METRIC THREADS

A modification of the foregoing method allows of threading to the base of a blind bore or to shouldered work without fear of an over-run when cutting SHORT metric threads with an English leadscrew, or English threads with a metric leadscrew. For reasons which will be understood in due course, the total carriage travel must not exceed the length capacity of the indicator being used. For right-hand threads the method is as follows:

Cutting passes are made with an inverted tool with the lathe spindle running in reverse. (Internal threading tools need a right-hand crank). A left-hand dead stop limits left-hand, hand traverse of the carriage and is set with the cutting tool at the desired inner starting position. The lathe is run backwards until the LSI reads 1, and the half-nuts are engaged. On completion of the cut, the lathe is stopped and the half-nuts disengaged. The tool is cleared, and the carriage hand traversed against the left stop. The lathe spindle is now run FORWARDS until the LSI again reads 1, which clearly shows that the work, leadscrew and carriage are all in the same relative positions as for the first cutting pass, whereupon the lathe is stopped, the half-nuts re-engaged and the lathe again started IN REVERSE for the next cutting pass.

An actual test of this method showed its practicability, but some care had to be taken to stop the lathe spindle at the right moment after its forward run to reset the LSI. A little previous experimenting will show the way in which backlash affects the LSI reading, this can then be allowed for, otherwise backlash may give a false indication, and attempts at engaging the half-nuts may force the carriage against the left stop, and perhaps move it.

The work length limits of 2 and 4 inches for corresponding indicators apply because of the confusion which would arise if the indicator was called upon to make more than one revolution. In this respect, too, it is important to note that if, with a 2-inch indicator, the total carriage travel is 2 inches and the cut is commenced at indicator reading 1, when the lathe is stopped on completion of the left to right cutting pass and the carriage is returned to its left stop the LSI will again read 1, showing one whole indicator turn. Pick-up will not be correct, however, until the lathe has been run *forwards* for a sufficient time for the indicator to make one revolution, 1 − 1. Similar remarks, of course, apply to a 4-inch indicator when the thread plus tool starting clearance requires a 4 inch carriage travel.

If the lathe has a metric leadscrew the LSI may read 30-60-90-120 and show a maximum carriage travel of 120 mm (4.7244 inches), but, or course, the routine will be the same as for the English leadscrew: after each cutting pass in reverse the lathe is run forwards until the LSI again reads the chosen starting figure.

This left to right working is suitable for all threads including odd pitches given by approximation gearing.

For left-hand threads the lathe spindle would run forwards for the cutting passes and in reverse while the half-nuts are disengaged and the leadscrew is driving the LSI back to its starting reading.

INDICATOR FOR METRIC LEADSCREW

An example of a metric leadscrew indicator is given in the diagram, Fig. 13 which shows the design fitted to a Harrison lathe with a metric leadscrew of 6.0 mm pitch. One revolution of the indicator dial registers a carriage travel of 120 mm, or 20 leadscrew revolutions.

The four numbered graduations indicate actual travel in millimetres, and each of the 20 subdivisions one leadscrew thread, or 6.0 mm of carriage travel.

The Instruction Plate reads:

A. 0.5, 0.75, 1.0, 1.5, 2, 3, 6 mm. pitch
Any line on indicator.
(i.e. engage half-nuts at random on leadscrew)

B. 1.25, 2.5, 5, 10 mm pitch
Any numbered line.
(i.e. groups of 5 leadscrew threads)

C. 4.0 mm pitch
Lines 60 or 120.
(i.e. any group of 10 leadscrew threads)

The pitches in set *A* are all exact sub-multiples of the leadscrew pitch, for which the half-nuts can be engaged at any convenient moment. The 6.0 mm leadscrew pitch holds exactly 12 0.5 mm threads, 8 0.75 mm threads, 6 1.0 mm threads, and so on.

Of set *B*, each pitch to be cut contains a whole number of threads in five leadscrew threads, one quarter turn of the indicator dial, 5 LST = 30 mm, holding 24

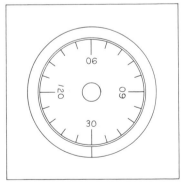

Fig. 13. A leadscrew indicator dial for metric leadscrews.

1.25 mm pitches, 12 2.5 mm pitches, 6 5.0 mm and 3 10.0 mm pitches.

Lastly, set *C*, the 4.0 mm pitch, shows 15 whole threads in 10 leadscrew threads. Note: although minimum pick-up for a pitch of 4.0 mm with a 6.0 mm pitch leadscrew is 2 LST to 3 component thread turns, it would be impracticable to read the unnumbered basic indicator graduations in pairs.

We may also note that 0.4 mm and 0.6 mm pitches will pick-up at any line on the indicator because 6.0 mm holds 15 whole 0.4 mm pitches and 10 whole 0.6 mm pitches. For thread pitches showing a whole number of thread turns only in groups of 3 or 7 leadscrew threads such as 0.45 mm pitch, holding 40 whole turns in 3 leadscrew threads, 1.75 mm pitch, holding 24 whole turns in 7 leadscrew threads, or 3.5 mm pitch, holding 12 whole turns in 7 leadscrew threads, the lathe must be reversed with the half-nuts remaining in engagement with the leadscrew until completion of the thread, or any of the other means previously outlined may be adopted to hold pick-up. This applies also to the cutting of threads/inch, DP or module worms and other pitches outside the scope of the leadscrew indicator.

It seems therefore that while an English leadscrew indicator will cover all ordinary English pitches in common use, similar facilities cannot be offered by a single indicator and a metric leadscrew for all common metric pitches.

REPEAT PICK-UP FROM 'CHALK MARKS'

Although today this method is never adopted, it was used extensively in the early days. Indeed, apart from the complete reversal method or the 'stop and

Fig. 14. An elementary diagram illustrating the principles of single-tooth dog-clutch control giving instant repeat pick-up for all thread pitches.

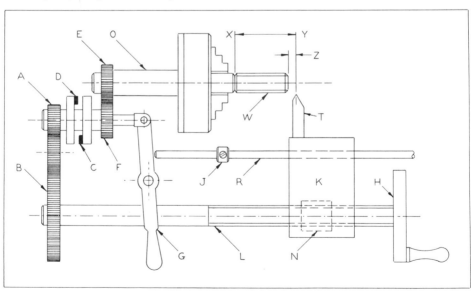

reset' routine, chalk marking was the only way of achieving repeat pick-up for threads that were not exact multiples of the leadscrew threads. The leadscrew indicator was invented long after the screwcutting lathe.

If, after the first threading pass, the carriage, leadscrew, lathe spindle and component can be restored to the exact relative positions each held at the commencement of the first pass, then on re-engaging the half-nuts, the second cutting pass will exactly follow the first.

Chalk-marking may therefore be carried out in the following way. With the lathe geared for the screw to be cut, the carriage is positioned so that the cutting tool is just clear of the starting end of the workpiece, and the half-nuts are engaged. Backlash in the gearing is taken up by rotating the lathe spindle through part of a turn. The tailstock body (e.g.) is then brought up to the carriage to form a right-hand dead stop. A chalk mark is now made on top of the leadscrew, and another mark on top of the chuck or work driving plate.

After the first threading pass, the lathe is stopped or slowed, the half nuts disengaged, and the carriage returned to the right-hand dead stop. The lathe is now slowly run until both the leadscrew and chuck chalk-marks take up their original positions, whereupon the half-nuts may be re-engaged and the second cutting pass made.

A practical test clearly demonstrated the disadvantages of this method. The best that could be done was to hand-pull the lathe driving belt, and at each chalk-marked revolution of the chuck, check for coincidence of the mark on the leadscrew: an appallingly slow and inefficient process. It is doubtful too, whether this approach could be relied upon except under those conditions where the relation-ship between component thread turns and leadscrew are simple, such as when cutting standard threads in the same language as that of the leadscrew.

DOG-CLUTCH CONTROL. AUTOMATIC PICK-UP.

We now turn to lathes of specialised design which give repeat pick-up without call for any special intervention by the operator.

As will be seen, the dog-clutch method is a modification of the complete reversal way of achieving pick-up, but it is generally necessary to purchase lathes with the refinement already fitted. The writer made and fitted the clutch to his lathe in 1973.

The accompanying diagram, Fig 14, illustrates the principle of operation, and shows that in addition to the normal screwcutting arrangements, a dog-clutch DC is interposed between the first quadrant driving gear A and the tumbler-reverse output spigot, the tumbler-reverse being represented by gears E and F with a 1-1 ratio. The clutch is of the 'single tooth' type: that is to say that each clutch member has only one dog, pin, or sector, so that when engaged, the angular relationship between each member, and hence between the headstock spindle and gear A, cannot change.

CUTTING PASS

For introductory purposes it will be necessary to assume that screwcutting will be carried out with the half-nuts remaining in engagement until completion of the thread.

Referring to Fig. 14, and starting from the position shown, to initiate a screwcutting pass, the half-nuts N are engaged, and the lathe spindle set in motion. Now the clutch $C\,D$ is engaged by means of the hand-lever G, which of course sets the leadscrew L in motion, thus traversing the carriage K and tool T from right to left (Y to X).

On near completion of the threading pass, the carriage K contacts collar J (previously suitably positioned and locked to rod R) thus moving rod R to the left where it abuts lever G and pushes the dogs out of engagement, which of course stops the leadscrew and arrests saddle traverse motion with the tool T at the X runout position. (Residual momentum in the leadscrew and its driving gears ensures that dog D fully disengages). The lathe spindle and workpiece may remain in forward motion.

CARRIAGE RETURN

To reposition the carriage and tool at Y ready for the next cutting pass, the tool is withdrawn, and the leadscrew rotated by means of the handwheel H, whereupon, after attending to tool depthing, the dog clutch may be re-engaged for the next cutting pass.

PICK-UP HELD

Repeat pick-up is assured because on turning the handwheel H to return the carriage and tool to the Y position, the leadscrew gear B drives gear A and the clutch member D through a series of pitch-by-pitch component thread turns without disturbing the relationship between the leadscrew and gears A and B. Also, with the half-nuts in engagement, at each single revolution of dog D, the leadscrew will move the carriage by exactly one component thread pitch distance in the direction X to Y, even though the leadscrew itself may make only part of a turn at each exact turn of dog member D.

Consequently, upon re-engaging the clutch for the next cutting pass, the moment dog C contacts dog D, all conditions will be exactly the same as for the first cutting pass, quite regardless of the ratio between gears A and B. We may also note that as the relationship between dog C and dog D can never be more or less than one exact component thread turn out of phase, and one component thread turn exactly equals one component thread pitch, there is no way in which correct relationships can be lost.

On repositioning the carriage to the Y position, the starting clearance Z can be held to any convenient minimum. Special applications of the clutch control that call for repeated specific settings for the X − Y length will be dealt with in due course.

LEFT-HAND THREADING

For cutting left-hand threads, rod R is locked clear of lever G, collar J is set and locked to rod R to locate the carriage at the X position, the leadscrew is reversed, and cutting passes are made from left to right, and when the tool has traversed clear of the workpiece, the clutch is disengaged by hand operating the lever G. After which, of course, the tool is repositioned at X by turning the hand-wheel H.

DOG-CLUTCH − ADVANTAGES

The advantages of a dog-clutch control for a leadscrew drive are overwhelming. Pick-up is assured for all pitches, regardless of the language of the leadscrew, and regardless of the pitch of the thread to be cut. Indeed it is impossible to devise any quadrant gearing with which pick-up would fail to hold.

Pick-up is always ready *immediately* a tool has been re-positioned for a fresh cutting pass, consequently no time is lost in waiting for a leadscrew indicator to register. With the clutch control a lead-screw indicator is unnecessary anyway.

The automatic arrest of traverse motion eliminates all fear of over-runs into the base of blind bores, or into shoulders, hence operator fatigue is greatly reduced.

With automatic arrest of traverse motion, much higher threading speeds are possible (up to 500 rpm in brass with the author's lathe, and up to 1000 rpm with the Hardinge).

Thread lengths can be pre-determined to within 0.001 inch (0.025 mm), and it is not necessary to pre-machine a runout groove: this is progressively formed by the threading tool, and takes the form of an annular recess having the same depth and form as that of the thread being cut.

The question of engaging half-nuts on to a fast revolving leadscrew never arises — all half-nut engagements are made on a 'dead' leadscrew, consequently all threading passes can be made without false starts.

A workpiece being threaded may remain in rotation for any length of time between cutting passes without loss of pick-up: a feature of value for the removal of thread crest burrs and cleaning prior to taking a measurement or testing the fit of a gauge. With the more common type of screwcutting lathe not fitted with the clutch, when it is necessary to thread with the half-nuts engaged until completion of a thread, (the complete reversal method), the time during which a workpiece can be allowed to rotate between cutting passes is strictly limited by the distance through which the carriage can travel along the lathe bed: an infernal nuisance.

As will be explained, the dog-clutch control permits the automatic indexing of the starts of a worthwhile range of multiple-start screw & nut threads *without stopping the lathe spindle and workpiece until completion of all starts to final depthings*, and at the higher threading speeds offered by the clutch. Or, with

temporary leadscrews of special lead, will similarly automatically index the starts of *any* multiple-start thread the lead of which is within the capacity of any given lathe without excessive stresses on the gearing: the lead limit may be taken, generally, as being twice that of the leadscrew: $\frac{1}{4}$ in. lead (4 tpi) for a lathe with a leadscrew of 8 tpi, and 6.0 mm lead for a lathe with a leadscrew of 3,0 mm lead, pro rata (see Section 6).

POWER LEADSCREW REVERSAL

The few industrial lathes at present fitted with the dog-clutch also have power reversal for the leadscrew (not to be confused with a tumbler reverse), the drive being taken from an additional but oppositely rotating clutch plate, and a handwheel is unnecessary. The diagram Fig. 15 shows one form of reversible clutch which, to the best of the writer's knowledge, was devised by the makers of the Hendey-Norton lathe circa 1914. That this clutch was not adopted as standard by all lathes can only be explained by assuming its action was not properly understood by other lathe manufacturers.

The following description assumes that threading operations will be carried out with the half-nuts N remaining in engagement with the leadscrew until completion of the thread being cut.

Referring to Fig. 15, the lathe spindle O drives bevel gear 1 (one) through gearing G – H at a 1 to 1 ratio. Bevel 2 is driven through bevel 3 in a direction opposite to bevel 1.

Bevel gears 1 and 2 are free to rotate on shaft 4 which carries a first-gear driver A for a leadscrew gear train, or for the input drive to a selective threading gearbox.

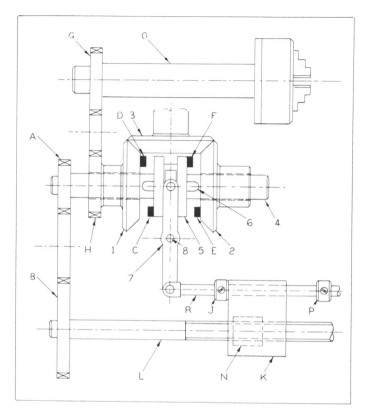

Fig. 15. An elementary diagram illustrating the principles of a reversible single-tooth dog-clutch control giving instant repeat pick-up for all thread pitches. Please also see text.

Bevel *1* is integral with driven spur gear *H*, and of course bevels *1* and *2* are restrained from endwise movement.

Boss *5* is coupled to shaft *4* by means of splines, one of which is shown at *6*, and upon which *5* is free to slide axially by movement of lever *7*, pivoted at *8*.

Bevels *1* and *2* each carry a single driving dog, *D* and *E*. Boss *5* carries driven dogs *C* and *F*.

OPERATION

Assuming the lathe carriage *K* is positioned clear of collar *J*, an anticlockwise movement of lever *7* slides boss *5* to the left, thus engaging dogs *C* and *D* which, with the lathe spindle rotating, sets shaft *4* and gear *A* in motion, thus driving the leadscrew *L*, which, with the half-nuts engaged, traverses the carriage *K* from right to left, where, on contacting collar *J* on the stop, or control rod *R*, lever *7* is moved to neutral, thus disengaging dogs *D – C* and arresting leadscrew rotation and carriage traverse. (The position shown in the diagram).

To reposition the carriage *K* to the right for the next cutting pass, lever *7* is moved in a clockwise direction, thus engaging dogs *E* and *F*. Shaft *4* and the leadscrew are therefore reversed, and the carriage K is driven up to collar P which pulls rod R to the right and moves boss 5 back to neutral, thus stopping the leadscrew and arresting carriage traverse.

Control rod R, of course, extends to the length of the lathe bed, and collars J and P are preset according to the position and length of a thread to be cut, plus a starting clearance.

LEFT-HAND THREADS

For cutting left-hand threads, the left-to-right carriage repositioning movement is used for cutting passes, and the right-to-left movement for non-cutting return passes.

NOTE: (1). When this type of reversible clutch arrangement is in use, the lathe spindle must not be run in reverse. A reversed lathe spindle reverses the action of lever 7 and stop collars J and P which, therefore, instead of releasing the clutch would jam it harder into engagement.

NOTE: (2). For returning the carriage to a right-hand dead stop by rack and pinion traverse for fast tool repositioning during screwcutting (q.v.) a reversible dog-clutch would require a handlever lock or stop to prevent inadvertent engagement of the reversing dogs.

APRON CONTROL

On some lathes, lever 7 (Fig. *15*) can be operated and the dog-clutch engaged (or hand disengaged) by means of a hand-lever mounted on the lathe apron and thus travelling with the carriage. This is convenient, indeed necessary, when cutting a thread at the right-hand end of a long shaft in a large lathe whereon a headstock mounted control lever would be out of reach. For rod R to serve this dual purpose, the rod probably has a keyway throughout its entire length, along which a handle-lever boss and key can freely slide

with traverse movement of the carriage. For operation of the clutch, this hand-lever would be used to rotate the stop rod through about 20 deg. either side of 'neutral'. Exact details are not available.

FAST RESETTING

Ordinarily, with power leadscrew reversal, non-cutting return passes take exactly the same time as the cutting passes: a feature to which objection is sometimes taken because of the time that can be devoted to repositioning a carriage by power when cutting long screws. In this respect it seems not to be generally understood that when cutting any standard thread in the same language as that of the leadscrew (and sometimes with conflicting languages) the non-cutting return passes *can* be made by rack-and-pinion traverse:

1. On completion of the first cutting pass (when the lead-screw will have automatically ceased to rotate) disengage the half-nuts. The workpiece may remain in motion.

2. By rack-and-pinion traverse, run the carriage to the right (X to Y, Fig. 14) through one, or as many SD units as may be necessary to position the tool clear of the workpiece to allow for a starting clearance. Re-engage the half-nuts. Set a right-hand carriage dead-stop.

3. Attend to tool depthing, and engage the dog-clutch for the next cutting pass.

4. On completion of the second cutting pass, disengage the half-nuts, run the carriage up to the right-hand dead-stop, and re-engage the half-nuts ready for the third cutting pass.

NOTE: When this method is being used, every cutting pass, once started, must be made in full, right up to clutch disengagement, otherwise correct pick-up will be lost.

RACK RESETTING EXAMPLE

Suppose a $10\frac{1}{2}$ in. length of 18 tpi thread has to be cut from a leadscrew of 4 tpi. For simplicity, let the material and diameter of the workpiece be such that threading has to be carried out at 200 rpm.

The number of 18 tpi thread turns in $10\frac{1}{2}$ in. is 189, which to traverse at 200 rpm would take 56.7 seconds, or, with a starting clearance, about one minute.

Let 10 be the number of threading passes required to complete the thread. If power dog-clutch reverse is used to reposition the carriage (X to Y, Fig. *14*) for each fresh cutting pass, then, of course, each return pass would take one minute and the total non-cutting return-pass time per screw would be 10 minutes.

Time Saved

Rack-and-pinion repositioning may take, say, three seconds, so the total time saved per return pass will be one minute minus 3 seconds = 57 seconds, or $9\frac{1}{2}$ minutes for ten return passes, therefore if 250 such lengths of thread were required, the time saved by rack-and-pinion resetting would be about one working week of 40 hours.

Finding the 'Y' Setting

For a lathe with a leadscrew of 4 tpi to cut 18 tpi, the basic gear ratio is 4/18 = 2/9, not further reducable in whole number terms, accordingly each 2 leadscrew threads hold exactly 9 component thread turns = $\frac{1}{2}$ inch SD. Obviously, $10\frac{1}{2}$ in. contain $\frac{1}{2}$ in. 21 times, so 22 SD units: X − Y = 11 in. would give a $\frac{1}{2}$ in. starting clearance, or, if the component has a pronounced chamfer, an X − Y distance of $10\frac{1}{2}$ in. would serve. As a matter of fact, finding the SD's, and hence the 'Y' position for all standard threads/inch to be

cut from an English leadscrew of 4 or 8 tpi (and with correct gearing as distinct from approximation gearing) is merely a matter of making a quick mental divisibility test: tpi not divisible by 2 = 1 in. SD, tpi divisible by 2 only = $\frac{1}{2}$ in. SD, tpi divisible by 2 and 4 = $\frac{1}{4}$ in. SD, and TPI the same as or an exact multiple of the leadscrew tpi have SD's the same as the leadscrew pitch: i.e. no 'wrong' positions on the leadscrew.

RACK RESETTING. METRIC.

Suppose a 250 mm length of 1.75 mm pitch has to be cut from a leadscrew of 6.0 mm pitch, at what distance should Y be set from X?

The basic ratio for 1.75 mm pitch with a leadscrew of 6.0 mm pitch is 1.75 to 6. Multiply by 4 to bring to whole numbers = 7/24 = 7 leadscrew threads to each 24 component thread turns minimum. 7 leadscrew threads of 6.0 mm pitch = 42 mm SD.

A thread length of 250 mm contains 42 mm about 5.9 times, but as we cannot have a fraction of an SD unit, we must set Y at 6 SD units from X, and 6 X 42 = 252 mm. This would allow a 2.0 mm starting clearance, which, with a chamfer, should be sufficient. If not, then one more complete SD unit would have to be added to the resetting X − Y distance, making X − Y = 294 mm, giving a starting clearance of 44 mm (about $1\frac{3}{4}$ in. − for which it may be necessary to use a half-centre support). However, even with a 44 mm starting clearance to be re-traversed at every cutting pass, there would obviously be a great saving in time by rack-and-pinion resetting.

RACK RESETTING. A TRANSLATION EXAMPLE

On one occasion the writer had to cut a $7\frac{3}{4}$ in. length of 2.0 mm pitch (ISO form) on a piece of silver steel of $\frac{3}{8}$ in. diameter. The leadscrew was of 8 tpi, geared thus:

$$35 - A - 25$$
$$36 - 80$$

which used No. 15 approximation for 50/127 (See Section 3). Accordingly the irreducable basic ratio was 35/25 × 36/80 = 63/100, showing exactly 63 leadscrew threads to 100 (nominal) 2.0 mm pitch thread turns, and 63 leadscrew threads of $\frac{1}{8}$ in. pitch = 7.875 in. Therefore, on completion of each cutting pass at X, the carriage was rack-and-pinion traversed to a right stop at Y $7\frac{7}{8}$ in. from X, which of course gave a $\frac{1}{8}$ in. starting clearnce. If memory serves, this thread was cut at about 200 rpm, each threading pass taking about 30 seconds. As very many passes were required to produce a clean bright thread, the time saved by rack-and-pinion resetting must have been between 20 and 30 minutes. (A travelling steady was not used, hence more and lighter cutting passes had to be made).

RACK RESETTING. ORDINARY LATHES (STOP AND RESET METHODS)

On ordinary lathes without a dog-clutch offering repeat pick-up and auto runout stop, when cutting threads for which a leadscrew indicator is of no use, such as 1.75 mm pitch from a leadsrew of 6.0 mm pitch, where the relationship is 7 leadscrew threads to 24 component thread turns, one is customarily advised to reposition the carriage for each fresh cutting pass by reversing both lathe spindle and leadscrew, with the half-nuts remaining in engagement until completion of the thread: the 'complete reversal method'.

But if one has to stop at the X runout position to reverse a lathe, one may as welll ascertain the SD or necessary $X - Y$

traverse, and instead of reversing the lathe, merely allow the whole to remain idle, disengage the half-nuts, and rack-and-pinion traverse up to a right-hand Y stop and re-engage the half-nuts, then restart the lathe spindle. This could still save 40 hours on a batch of long screws: more in fact, because without automatic leadscrew stop at X, threading speeds have to be much slower, and hence return passes with the half-nuts engaged would take longer, even twice as long, thus showing a possible saving of two weeks on a batch.

However, rack-and-opinion carriage resetting on an ordinary lathe, which the writer has personally tested and found entirely practicable, is greatly aided by:

1. A runout groove or recess equal in width to about $1\frac{1}{2}$ to 2 component thread pitches, to allow for slight variations in the X runout position, and,
2. A 'buffer' or flexible type left-hand runout indicator (as distinct from a left-hand carriage dead-stop) the flexibility in the left indicator 'stop' being to allow for the inevitable slight carriage overruns at the X runout position.

The diagram, Fig. 16 illustrates the requirements. An extendable indicator rod R is free to slide axially in a bearer D (preferably mounted on the lathe headstock). Rod R is provided with two axially adjustable collars, A and B with a compression spring S interposed between collar B and the fixed bearer D. With this arrangement, slight over-runs of the carriage K merely compress spring S, consequently there is no jamming which would prevent half-nut disengagement. When the carriage is returned to the right-hand dead stop C after a slight over-run at X, the spring S returns the indicator rod R to its original position, with collar A abutting the bearer D.

Initial settings may be exactly as shown

in the diagram with the right-hand stop C positioned about one-quarter of a leadscrew pitch (1.5 mm for a leadscrew of 6.0 mm pitch) away from the carriage K when this is in the Y position. This small gap G allows for the occasional early stopping of the lathe under which circumstances the leadscrew would not have revolved quite enough to permit re-engagement of the half-nuts at the Y position. Before the half-nuts can be engaged on a wrong leadscrew thread at Y, an over-run at X has to amount to one whole leadscrew thread pitch, which is unlikely. On the other hand, if a stop at X is made exceptionally early, thus preventing re-engagement of the half-nuts at Y, then it is permissible to rotate the chuck by hand through a sufficient portion of a revolution to bring the leadscrew threads into a favourable position for half-nut re-engagement. As a matter of fact, when testing this method, four complete threads were cut from a leadscrew of 8 tpi, all with awkward pick-up of SD figures, the minimum possible X − Y settings being 1 in., $1\frac{1}{8}$ in., $1\frac{1}{2}$ in. and $3\frac{1}{4}$. three of these threads being of metric pitch. On no occasion was it possible to re'engage the half-nuts on to a wrong set of leadscrew threads despite the fact that very many threading passes were taken.

During threading it is advisable to watch the gap between the carriage and the projecting end of the indicator rod: indeed this is necessary when internally threading blind bores by normal right to left traverse on any lathe without a dog-clutch.

Any doubt over whether or not a right-hand stop is correctly positioned is easily dispelled by taking a threading pass with the tool slightly withdrawn, and observing that the tool is in fact following the first helix cut.

GENERAL DOG-CLUTCH NOTES

The author's dog-clutch may be seen in the photograph Fig. 17, and Fig. 17A shows the clutch mechanism in section. This clutch is engaged against the compression of a strong spring, and is held in engagement by an over-centre lever linkage. Collapse of this linkage and instant clutch disengagement is effected by carriage traverse motion abutting a stop-rod, or by means of the hand control

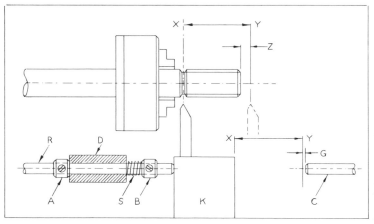

Fig. 16. Illustrating the principles of a 'buffer type' thread runout indicator for rack resetting on lathes without dog-clutch control for the leadscrew.

Fig. 17. The author's single-tooth dog-clutch mechanism. The crosshead assembly does not rotate.

KEY TO PARTS in the sectional view below

A. Main spigot or support shaft.
B. 25T pinion (driven by tumbler reverse)
C. Driving dog boss.
D. 40T driver for gear trains.
E. Sliding-dog boss, free on A.
F. Free collar.
G. Retaining collar.
H. Retaining washer.
K. Pull-off (disengaging) spring.

lever when necessary. This instant disengagement ensures that the dogs do not drive on their tips on near completion of threading passes. At the time of writing, this clutch has operated over eighty thousand times with full reliability.

DOG-CLUTCH RATIOS

Although Figs. 14 and 15 show a 1 to 1 ratio between the lathe spindle and dog-clutch, pick-up can be held with 1 to 2, 1 to 3 etc reduction gearing between the lathe spindle and the clutch, correct pick-

I. Coupling rods.
M. Crosshead link.
N. Adjusting nut.
O. Sliding dog.
P. Non-sliding dog.
R. Operating rod (linked to hand-lever control and trip mechanism)
T. Space for tumbler-reverse lever pivot.

On moving rod R to the right, against the pull of spring K, dog O is brought into engagement with dog P, and gear D is set in motion.

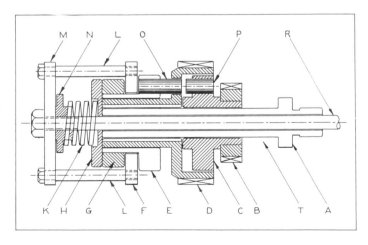

up moments then occur at every second, third etc revolution of the lathe spindle, and of course, at the lower clutch speeds the shock at the moment of engagement is correspondingly reduced.

An Italian Patent (the Manaioni) goes so far as to advocate the use of leadscrews of 1 inch lead (or 25.0 mm lead) i.e. 1 tpi, and a 1 to 8 reduction between the lathe spindle and the dog-clutch with the idea of threading at speeds of up to 2000 rpm, at which speed clutch engagement would be at only 250 rpm, although one would imagine that the advantages for short threads would be minimal, if any.

This Patent also stated that if the reduction gearing between the lathe spindle and dog-clutch is temporarily modified, for example, to 2 to 3, pick-up would occur at every $1\frac{1}{2}$ revolutions of a piece being threaded, thus making it possible to automatically index the starts of a 2-start thread. The writer took the trouble of testing this assertion by temporarily removing the tumbler-reverse through which his dog-clutch is driven at a 1 to 1 ratio, and substituting reduction gearing in the ratio 2 to 3 with special gears of 36 and 54 teeth. An attempt was then made to automatically index the starts of a 2-start thread of $\frac{1}{8}$ in. lead, $\frac{1}{16}$ in. pitch with a leadscrew of 8 tpi — something that ordinarily cannot be done. The lathe was geared:

$$36 - 54$$
$$CL$$
$$40 - 65 - 30$$
$$45 - 40$$

where CL symbolises the single-tooth dog-clutch. The overall ratio between the lathe spindle (the 36) and the leadscrew (the last 40) was 1 to 1 for the $\frac{1}{8}$ in. lead.

Two starts of $\frac{1}{16}$ in. pitch were certainly obtained, but as there was no guarantee as to which of the two helices would pick-up at any single cutting pass, one helix

would be repeatedly and increasingly depthed with the result that when, by chance, the second helix came into phase, the extra heavy cut would tear the flanks and spoil the thread. A similar experiment with the input gears reversed to 3 to 2 (54 to 36) and the remainder of the gearing readjusted to give a lead of $\frac{1}{8}$ in. was tried with the object of cutting a 3-start thread of $\frac{1}{8}$ in. lead and 0.04166 in. pitch. Again, three starts were obtained, but with the same unreliability over which of the helices would pick-up at any given cutting pass, so unless there is some additional feature not disclosed, the method must be written off as of no practical use. The special 36 and 54T gears that had to be made for this experiment were subsequently re-bushed for use with the standard change gear set.

REPEAT PICK-UP: JAPANESE

These notes would be incomplete without mention of a Japanese lathe whereon the leadscrew nut is in the form of recirculating bearing-balls permanently engaged with the leadscrew. When screwcutting, the leadscrew nut is prevented from rotating by means of a pin or lock, and users are advised to cut all threads by the complete reversal method: despite the fact that the lathe is driven by a motor of about 7 horse-power which, at the moment of each reversal, must place a severe loading on the electricity supply.

Ordinary traverse of the carriage by hand is probably carried out by rotating the leadscrew nut along the leadscrew by means of a hand wheel and bevel gears on the apron.

Not having an opportunity to experiment with this type of lathe, it is not possible to offer any hints that might hasten carriage repositioning — except perhaps that of temporarily speeding up the lathe spindle for non-cutting return passes, which seems to be a somewhat clumsy approach, with the risk that one might forget to slow down again for each cutting pass, even assuming that the lathe could be speeded up and slowed down quickly enough to make it worthwhile.

PICK-UP AND APPROX TRANSLATION RATIOS

As already mentioned, the lower the error in any approximate ratio for 50/127 or vice-versa, the greater the minimum SD or X − Y pick-up distance.

This of course has little effect on times for the more common short thread lengths when threading with a reversible single-tooth dog-clutch, or by the complete reversal method, but low error approximations generally entirely preclude all possibility of rack-and-pinion carriage resetting between cutting passes when cutting threads of any significant length.

For example, the Colchester Lathe Co's approximation of 6/7 × 8/11 × 12/19 which approximates 50/127 with an error of plus one part in 36575 parts (or about plus 1.7 in. a mile) when used in conjunction with a leadscrew of 4 tpi to cut a thread of 6.0 mm pitch, shows a minimum SD (or X − Y) figure of 6912 leadscrew threads = 144 feet: nearly 29 5-foot lathebed lengths. This also means, incidentally, that if X − Y happened to equal in length 48 leadscrew threads (12 inches) and the lathe was stopped at X, and the carriage rack traversed to Y, then on re-starting the lathe, one would have to wait until the leadscrew had made 6912 minus 48= 6864 revolutions before the half-nuts could be re-engaged for the next cutting pass. Consequently, if, for example, a significant number of 12 inch lengths of 6.0 mm pitch thread were required, and a lathe with a metric leadscrew was not available, considerable time could be saved by ignoring the built-

in translation ratio, and setting another on the quadrant, with a more favourable (i.e. much shorter) pick-up, or X − Y (Fig. 14) distance suitable for rack resetting between cutting passes.

Admitted, in general, the shorter the pick-up distance, the less accurate the translation ratio, but on the other hand it is generally only surmise that a leadscrew is of perfect lead, and if a thread was required with a specified very small lead or pitch error it would be most unwise to attempt to cut it on an ordinary centre lathe, no matter how accurate the leadscrew and gearing may be thought to be, simply because, even after the greatest care, if a customer rejected such a screw as having a pitch outside the limits laid down, the time devoted to making the screw would be entirely wasted, with no possibility of argument.

Accordingly, if small pitch errors are acceptable, screw-cutting can be considerably speeded up by selecting a suitable ratio from the accompanying Tables, T10A and T10B which offer a range of basic ratios for 30 metric pitches or leads to be cut from a leadscrew of 4 tpi.

NOTE: For a leadscrew of 8 tpi, halve the denominator. If this is not possible, then let the denominator stand, and double the numerator. Maximum pitch for a leadscrew of 8 tpi should not exceed 6.0 mm. (If a selective gearbox is fitted the maximum metric pitch cut from a leadscrew of 8 tpi should not greatly exceed 3.0 mm).

The tables are also used for ascertaining the best approximate ratios for the automatic and semi-automatic indexing of the starts of multiple-start threads − q.v. − Section 6.

In the tables, examples with errors of less than one part in 1000 parts (0.001 in. per inch) are indicated thus *. No ratio

Pitch or Lead mm	Exact Basic Ratio	1	2	3	4	5	6	7	8	9	10	11	12
					Approximate Ratios						Basic Minimum		
12.0	240/127	17*/9	36/19	49/26	66/35	70/37	74/39	2×50/1×53	2×69*/1×73				
11.5	230/127	29*/16	38*/21	49/27	67*/37								
11.0	220/127	19/11	26*/15	33/19	45*/26								
10.5	210/127	28/17	33/20	38*/23	43*/26	48*/29	53/32	58/35	63/38	68/41	71/43	81*/49	
10.0	200/127	11/7	30/19	41/26	52*/33	63*/40	2×37*/1×47	2×40/1×51	2×48*/1×61	2×56/1×71	2×59*/1×75	2×67/1×85	2×70/1×89
9.5	190/127	3/2											
9.0	180/127	17*/12	24/17	27/19	44/31	2×39*/1×55	2×46/1×65	2×49/1×69					
8.5	170/127	4/3	47/35	51/38	55/41	59/44	63/47	67/50	71*/53	75*/56			
8.0	160/127	24/19	34*/27	39/31	44/35	63*/50	2×46*/1×73						
7.5	150/127	13*/11	20/17	32/27	33/28	45/38	46/39	58/49	59*/50	71/60	72*/61	84/71	85*/72
7.0	140/127	11/10	21/19	32*/29	43*/39	54*/49	76*/69	85/77					
6.5	130/127	38/37	39/38	40/39	41/40	42*/41	43*/42	44*/43	45*/44	46/45	47/46	81/79	83/81
6.0	120/127	17*/18	18/19	33/35	35/37	37/39	49/52	50/53	52*/55	53/56	67/71	69*/73	71/75
5.5	110/127	13*/15	19/22	33/38	45*/52								
5.0	100/127	11/14	15/19	26*/33	37*/47	40/51	41/52	48*/61	56/71	59*/75	63*/80	67/85	70/89
4.5	90/127	12/17	17*/24	22/31	27/38	29/41	39*/55	46/65	49/69	56*/79	61*/86		

Columns **1–7** fall under the heading **Approximate Ratios**; columns **8–12** fall under the heading **Basic Minimum**.

Pitch or Lead mm	Exact Basic Ratio	1	2	3	4	5	6	7	8	9	10	11	12
4.0	80/127	12/19	17*/27	22/35	39/62	46*/73	63*/100						
3.5	70/127	11/20	16*/29	21/38	27*/49	38*/69	43*/78						
3.0	60/127	17*/36	25/53	26*/55	33/70	35/74	37/78	1×49/2×52	1×53/2×56	1×67/2×71	1×69*/2×73	1×71*/2×75	
2.5	50/127	11/28	13*/33	15/38	20/51	24*/61	28/71		1×37*/2×47	1×41/2×52	1×59*/2×75	1×63*/2×80	1×67/2×85
2.0	40/127	6/19	11/35	17*/54	23*/73	1×39/2×62	1×63*/4×50						
1.75	35/127	8*/29	11/40	19*/69	1×21/2×38	1×27*/2×49	1×43/4×39	—					
1.5	30/127	13*/55	1×17*/2×36	1×25/2×53	1×33/2×70	1×35/2×74	1×37/4×39	1×49/4×52	1×53/4×56	1×67/4×71	1×69*/4×73	1×71/4×75	1×67/4×85
1.25	25/127	10/51	11/56	12*/61	13*/66	14/71	15/76	1×35/2×89	1×37*/4×47	1×41/4×52	1×59*/4×75	1×63*/4×80	1×67/4×85
1.0	20/127	3/19	11/70	14/89	1×17*/2×54	1×23*/2×73	1×26*/5×33	1×37*/5×47	1×39/4×62	1×63*/8×50			
0.8	16/127	1×12/5×19	1×17*/5×27	1×22/5×35	1×39/5×62	1×46*/5×73	1×63*/5×100						
0.75	15/127	1×13*/2×55	1×17*/4×36	1×25/4×53	1×33/4×70	1×35/4×74	1×37/8×39	1×49/8×52	1×53/8×56	1×67/8×71	1×69*/8×73	1×71/8×75	
0.7	14/127	1×13*/2×59	1×15*/4×34	1×16*/5×29	1×27*/5×46	1×41*/12×31	1×43*/10×39	1×44*/7×57					
0.6	12/127	5/53	7/74	1×17*/5×36	1×26*/5×55	1×33/5×70	1×37/5×78	1×49/10×52	1×53/10×56	1×67/10×71	1×69*/10×73	1×71/10×75	
0.5	10/127	1×11/2×70	1×13*/3×55	1×17*/6×36	1×23*/4×73	1×25/6×53	1×35/6×74	1×37/6×78	1×49/12×52	1×53/12×56	1×67/12×71	1×63*/16×50	1×71/12×75

Table T10. Giving basic gear ratios for metric pitches from a leadscrew of 4 threads/inch.

*Denotes ratios with an error of less than 0.1%.

None have an error of more than 0.4%.

To check the metric pitch given by the gearing shown:

$$P(\text{mm}) = \frac{\text{Driver(s)}}{\text{Driven}} \times \frac{127}{20}$$

For a leadscrew of 8 threads/inch the same ratios can be converted:-
(1) If possible, halve the denominator.
(2) If (1) is not possible, let the denominator stand, and double the numerator.
Note: The metric pitch cut from a leadscrew of 8 threads/inch should not exceed 6.0 mm, and this can severely strain a small lathe.

holds an error greater than one part in 250 parts (plus or minus 0.004 in. per inch).

The actual metric pitch given by any of the ratios can be found from: Driver/Driven × 6.35 (or × 127/20).

For example, a ratio of 17/18 for 6.0 mm pitch gives an actual pitch of 17/18 × 6.35 = 5.9972221 mm, an error of 0.00277 ... mm or minus 0.00011 in. with a perfect leadscrew.

All ratios in the tables are in basic form, i.e. not further reducable in integral number numerator and denominator terms, therefore, of course, many require resolving into suitable change gear form. It is assumed that the lathe has a selective screwcutting gear box, and that any necessary gears for 'outside' ratios will be obtained from the lathe makers.

The tables show ratios in basic form to facilitate the selection of minimum pick-up or X − Y settings for any metric pitch to be cut. In each example, the numerator equals the minimum possible number of leadscrew threads in any X − Y setting, therefore any numerator multiplied by the leadscrew pitch gives a measurement for minimum X − Y.

For example, taking the 17/18 ratio for a pitch of 6.0 mm: 17 leadscrew threads of 0.25 in. pitch = 4.25 in. minimum SD, or X − Y setting. Hence, according to the length of a 6.0 mm pitch to be threaded, the X − Y resetting distances may be made as follows:

SETTING No.	X–Y MINIMUM (Inch)	FOR THREAD LENGTHS UP TO (Inch)
1	4.25	4.125
2	8.5	8.375
3	12.75	12.625
4	17.0	16.875
5	21.25	21.125

In each example, a minimum component starting clearance of $\frac{1}{8}$ in. is allowed for. Thus, for a 12 in. length of 6.0 mm pitch, No. 3 setting would have to be used, the starting clearance (Z) being $\frac{3}{4}$ in.

We may note that the 17/18 ratio for threading 6.0 mm pitch is derived from No. 27 approximation for the 50/127 translation ratio, and with this approximation, the minimum X − Y resetting holds for the following pitches: 12.0, 9.0, 6.0, 4.0, 3.0, 2.0, 1.5, 1.0, 0.8, 0.75, 0.6 and 0.5 mm, although the approximation ratio with the 17 numerator will not necessarily always appear in column 1 of the tables.

USE OF TABLES

Some further examples of the use of the tables will now be given.

1. Required: a 3 in. length of 3.5 mm pitch. We have a choice of six approximate ratios, with numerators from 11 to 43. No. 1 shows 11 leadscrew threads to 20 component thread turns. 11 LST of 0.25 in. pitch span 2.75 in., hence X − Y would be too short. Two such X − Y units = 5.5 in. which would be inconveniently long with a Z starting clearance of $2\frac{1}{2}$ in. The next ratio, No. 2, 16/29 = 16 X 0.25 = 4 in. for X − Y which of course would cover the required 3 in. length with 1 in. Z clearance.

The actual pitch given by the 16/29 ratio = 3.5034482 mm: a pitch error of plus 0.0001357 in.

2. Required: a 5 in. length of 4.0 mm pitch. For this pitch, the table shows a range of six approx ratios. 5 in. = 20 leadscrew threads of 0.25 in. pitch. No. 3 approx 22/35 should serve. 22 LST span $5\frac{1}{2}$ in. = X − Y, thus allowing a comfortable $\frac{1}{2}$ in. Z clearance. The pitch given = 3.9914285 mm, the error being minus one part in 466.

3. Required: a 10 in. length of 7.0 mm

pitch. Here we have a choice of six approx ratios, four of which are starred. 10 in. = 40 leadscrew threads, No. 4 approx 43/39 would therefore serve. 43 LST of 0.25 in. pitch = 10.75 in., so, with X – Y = $10\frac{3}{4}$ in., we would have a $\frac{3}{4}$ in. Z clearance. The error with this approx is plus 1 in 5460 (No. 17 approx for 50/127).

For the finer pitches we should remember to scan the exact basic ratio columns. For example, minimum X – Y for 1.0 mm pitch geared 20/127 is 20 leadscrew threads = 5 in. Only Nos. 1, 2, 3 and 4 approx ratios give shorter X – Y rack resetting distances.

METRIC/ENGLISH RACK RESETTING

The Tables T11A & B give a selection of approximate ratios for cutting threads/inch from a metric leadscrew of 6.0 mm pitch.
NOTE: For a leadscrew of 3.0 mm pitch, halve the denominator. If this is not possible, let the denominator stand, and double the numerator. Maximum threads/inch for a leadscrew of 3.0 mm pitch should not exceed 4.

The metric/inch tables are used in exactly the same way as the inch/metric just explained, with the exception that the *exact basic* ratios cannot be used for any thread length much shorter than 127 leadscrew threads = 762 mm (30 in.) Two tpi: 2.75 and 3.25, offer a minimum pickup of 1524 mm and 2.625 and 2.875 tpi show a minimum pick-up of 508 LST = 3048 mm.

To check the inch pitch given by any approx ratio, multiply Driver/Driven by 30/127 (or by 0.2362204). The reciprocal will of course give the threads/inch.

For example, No. 2 approx for 8 tpi: 9/17 × 30/127 = 0.1250578 in. and the reciprocal = 7.9963025 tpi. The pitch error is plus 57.8 millionths of an inch.

Some examples of the Metric/Inch tables in use will now be given.
1. Required: a 10 in. length of 8 tpi.
  10 in. = 254 mm = 254/6 = 42.333 ... i.e. 43, 6.0 mm pitch leadscrew threads. No. 2 approx: 9/17 is starred. 43/9 = 4.77 ... 'Round' to 5 = 5 × 9 = 45 leadscrew threads of 6.0 mm pitch = 270 mm X – Y. The component thread length = 254 mm, hence the Z clearance will be 270-254 = 16 mm (about $\frac{5}{8}$ in.).
2. Required: a 12 in. length of 6 tpi.
  12 in. = 304.8 mm. Divide by 6 (the leadscrew pitch) = 50.8, 'round' to 51 leadscrew threads. Accordingly we require, if possible, a LST (numerator figure) of between 51 and 52 or so. We may note that No. 1 approx: 17/24 with numerator multiplied by 3 = 51 LST. 51 × 6 = 306 mm for X – Y which would allow a Z clearance of 1.2 mm (0.047 in.) which should serve with a fairly well pronounced starting chamfer. The pitch error with the 17/24 approx is plus 0.000656 in.

If we try the better approximation No. 2: 12/17 we find 4 × 12 = 48 LST which is too short, and 5 × 12 = 60 LST = 360 mm which for a component length of 304.8 mm would leave a Z gap of 55.2 mm (2.17 in.) which could be of inconvenient length.
3. Required: a 5 in. length of 3 tpi. 5 in. = 127 mm. 127/6 = 21.16 = 22 LST. The nearest approx ratio is No. 2: 24/17. 24 × 6 = 144 mm. X – Y. 144 mm minus component length of 127 mm = 17 mm (0.67 in.) Z clearance. The pitch error from this ratio is plus 154 millionths/inch.

Pitch or Lead mm	Exact Basic Ratio	Approximate Ratios								Basic Minimum			
		1	2	3	4	5	6	7	8	9	10	11	12
2	127/60	17/8	19/9	36*/17	53/25	55/26	70/33	74/35	87/41	89/42			
2.5	127/75	17/10	22*/13	27/16	39/23	56/33	100*/59						
2.625	508/315	21/13	29*/18	34/21	45/28	50*/31	66/41	71*/44					
2.75	254/165	17/11	20*/13	37/24	43/28	54/35	57*/37	63/41					
2.875	508/345	22/15	25/17	28*/19	31/21	47/32	53*/36	59/40	81*/55				
3	127/90	17/12	24*/17	31/22	38/27	41/29	58/41	65/46	75/53				
3.25	254/195	13/10	17/13	30/23	43*/33	47/36	56*/43	64/49	69*/53				
3.5	127/105	17/14	23*/19	29*/24	35/29	40/33	52*/43	63/52	64/53	75*/62	81*/67		
4	127/120	17/16	18*/17	19/18	35/33	37/35	53/50	55*/52		87/82	88/83	89/84	
4.5	127/135	16*/17	31/33	47*/50				63*/67	65/69	76/81	78/83	79*/84	2×55*/3×39
5	127/150	11*/13											
6	127/180	12*/17						1×75/2×53					
7	127/210	17/28	20/33	23/38	26*/43	29*/48	32/53	35/58	1×63/2×52	1×75*/2×62	1×81*/2×67		
8	127/240	9*/17	28*/53	37/70				44/83	53/100	1×55*/2×52	1×87/2×82	1×89/2×84	
9	127/270	8*/17	25*/53	31/66	33/70	38/81	39/83	47*/100	1×49/2×52	1×55*/3×39	1×63*/2×67	1×65/2×69	1×79/2×84
10	127/300	11*/26	17*/40	23/54	25*/59	27/64	39/92						
11	127/330	5*/13	17/44	23/60		1×43/2×56	1×57*/2×74	1×63/2×82					

Approximate Ratios columns 1–8; **Basic Minimum** columns 9–12.

Pitch or Lead mm	Exact Basic Ratio	1	2	3	4	5	6	7	8	9	10	11	12
12	127/360	6*/17	17/48	29/82	31/88	1×41/2×58	1×55*/4×39	1×65/4×46	1×67*/2×95	1×73*/9×23	1×75/4×53		
14	127/420	10/33	13*/43	16/53	17/56	23*/76	29*/96	1×35/2×58	1×63/4×52	1×75*/4×62	1×81*/4×67	1×89/4×84	
16	127/480	9*/34	13/49	17/64	19/72	22/83	1×35/2×66	1×37/2×70	1×53/4×50	1×55*/4×52	1×87/4×82		
18	127/540	4*/17	19/81	1×25/2×53	1×31/2×66	1×33/2×70	1×39/2×83	1×47*/4×50	1×49/4×52	1×55*/6×39	1×63*/4×67	1×65/4×69	1×79*/4×84
19	127/570	2/9	1×29/5×26	1×33*/4×37	1×37*/2×33	1×39*/5×35	1×41*/4×46	1×49*/4×55	1×76*/11×31				
20	127/600	7/33	11*/52	17/80	1×25*/2×59	1×27/2×64	1×39/2×92						
22	127/660	5*/26	17/88	1×23/2×60	1×27/2×70	1×43/4×56	1×57*/4×74	1×63/4×82					
24	127/720	3*/17	17/96	1×19/2×54	1×29/2×82	1×31/2×38	1×41/4×58	1×55*/8×39	1×65*/8×46	1×67*/4×95	1×73*/9×46	1×75/8×53	
26	127/780	7*/43	8/49	1×22*/5×27	1×29*/2×89	1×43*/4×66	1×57*/5×70	1×85*/6×87					
28	127/840	5/33	8/53	1×13*/2×43	1×13/2×56	1×23*/2×76	1×29*/2×96	1×35/4×58	1×44*/3×97	1×63/8×52	1×75*/8×62	1×81*/8×67	
32	127/960	9*/17	11/83	13/98	1×17/2×64	1×19/2×72	1×35	1×37/4×70	1×53/8×50	1×55*/8×52	1×87*/8×82		
36	127/1080	2*/17	1×19/2×81	1×25/4×53	1×31/4×66	1×33/4×70	1×39/4×83	1×47*/8×50	1×49/8×52	1×55*/12×39	1×63/8×67	1×65/8×69	1×79/8×84
40	127/1200	7/66	1×11*/2×52	1×17/2×80	1×25*/4×59	1×27/4×64	1×39/8×46						

Table T11. Giving basic gear ratios for threads/inch from a lead-screw of 6.0 mm. pitch.

*Denotes ratios with an error of less than 0.1%

None have an error of more than 0.4%.

To check the pitch given by inch measure from the gearing shown:

$$P_{(in.)} = \frac{\text{Driver(s)}}{\text{Driven}} \times \frac{30}{127}$$

For a leadscrew of 3.0 mm. pitch, the same ratios can be converted:-

(1) If possible halve the denominator.

(2) If (1) is not possible, let the denominator stand, and double the numerator.

Note: Threads/inch cut from a leadscrew of 3.0 mm pitch should not be less than 4 (four), and this can severely strain a small lathe.

SECTION 6

Multiple-start threads

If an 8 tpi helix is cut to 55 deg. Whit. form and instead of depthing to the full 0.080 in. the helix is depthed to one half that amount, i.e. 0.040 in., then very broad thread crests will remain, and in fact, these crests will be of exactly the necessary width to allow the cutting of a second tpi helix of 0.040 in. depth between the first helix.

The resulting threads will then have much of the appearance of 16 tpi, and in fact, a 'saw-tooth' type thread gauge with 16 'teeth' to the inch would fit nicely.

Such a 'double' thread would be known as a '2-start' thread. (see also Fig. 18) The means adopted to cause a threading tool to cut the second start or helix exactly mid-way between the first is known as *start indexing.* Any thread with 2 or more individual helices is known as a *multiple-start thread.*

Had the 8 tpi lead in our example been cut to a depth of $\frac{1}{3}$ of 0.080 in. = 0.02666 in., then the first 8 tpi lead helix would leave thread crests of sufficient width to fit in two additional helices, and the result would be a 3-start thread of 0.125 in. lead and 0.041666 in. pitch, and the whole would have much of the appearance of an ordinary thread of 24 tpi.

A workpiece may hold any number of individual helices (starts), but the most common would be 2, 3, 4, 5 and 6.

LEAD AND PITCH

Hitherto, and by long custom, we have indiscriminately referred to the *pitch* of ordinary single threads, and to the *pitch* of *lead*screws. We are entitled to do this because in a single thread, pitch carries the same meaning as lead. However, with multiple-start threads there is an important although simple distinction between lead and pitch, and much confusion can be avoided by noting this:

LEAD

The lead of a screw is the distance through which it would advance axially on being rotated one turn (360 deg.) through a fixed nut. This definition applies to any screw, single or multiple threaded.

PITCH

The pitch of any thread is the measured distance from a point on one thread to a corresponding point on the *next adjacent thread.* In other words, pitch is the distance from, e.g., the centre of one thread crest to the centre of the next *nearest* thread crest, and this applies to any thread, single or multiple threaded.

Hence in our example of an 8 tpi *lead,* 2-start thread, as soon as we cut the second helix midway between the first, we produced a *pitch* of $\frac{1}{16}$ in., even though the lathe remained geared to cut a *lead* of 0.125 in.

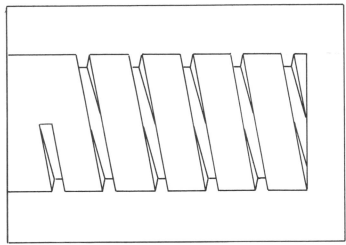

Fig. 18. MULTIPLE START THREADS. With the thread shown, there is obviously sufficient space between thread turns for the cutting of a second and similar thread. The screw would then be termed 'a two-start thread'.

Sometimes there is a tendency to regard the pitch of an English multiple start thread from the point of view of 'thread per inch', for example, 2-start, 0.125 in. lead, '16 tpi pitch'. This approach is however not to be recommended, and except insofar as in this example the notation would inform a turner to depth the helices to the same amount as for a thread of 16 tpi, the statement is ambiguous because although the threads would have every appearance of an ordinary single-lead 16 tpi, an ordinary 16 tpi nut would certainly not fit. It would however be quite in order, indeed sometimes convenient, to refer to the pitch as for example '16 *pitches per inch'*, abbreviated to 'ppi' – as distinct from tpi.

Fortunately, for metric multiple start threads there is less chance of confusion because these are designated entirely by lead and pitch, and there is no possibility of referring to the pitch in any other terms.

USES OF MULTIPLE THREADS

Multiple-start threads are used when it is necessary to produce a large axial, (endwise) movement from a small or modest rotary movement, such as to focus a camera lens, to operate the ram of a fly-press,or to release a lathe tool turret for indexing. Multiple-start threads may also often be seen on jam and pickle-jars to facilitate a rapid removal of the tops, although in these instances there is seldom a full turn of any one start.

But why not merely use a coarse thread? This is best answered by an example. Suppose you have a brass tube with a wall thickness of 0.070 in. and you wish to thread it to advance $\frac{1}{8}$ in. in one turn. This would ordinarily call for a thread of 8 tpi, which in the Whit. form has a depth of 0.080 in., so of course, in attempting to thread the tube you would break through into the bore, whereas, as you have seen in our opening example, you can cut two separate 0.125 in. leads to half depth, or three 0.125 in. leads to $\frac{1}{3}$ depth, the latter leaving a tube wall-thickness of 0.043 in.

STANDARDISATION

On searching both "Machinery's Handbook" (Ed. 17) and "Machinery's Screw Thread Book" (Ed.20) I can find no

reference to any standardisation for multiple-start threads, we are free to design multiple-start threads by:
  (1) Fixing the lead and number of starts, and accepting the resulting pitch, or
  (2) Fixing the pitch and number of starts, and accepting the resulting lead.

From the point of view that pitches have been standardised for ordinary single threads, the second option might appear to be more reasonable, yet if the axial advance per revolution is to be the ruling factor, then the resulting pitch may sometimes be non-standard. Of course, if you are working for someone else, then the decision will have been made for you unless you are the draughtsman.

Before we can cut a multiple-start thread we need to know the *lead* so that a lathe can be geared accordingly. The only reason we require a knowledge of the pitch is for the purpose of depth to the correct amount. Pitch depthings will be found under "Quick Reference Thread Information Summary" (Page 10 Section 1).

The following formulas will cover all ordinary requirements for multi-start threading.

1.   When the lead (expressed as tpi) is adopted as the reference, then the pitch will vary according to the number of starts:

$$\text{Pitch} = \frac{1}{\text{Lead tpi} \times \text{number of starts}}$$

1a.   If the lead is given in linear measure, and the pitch is preferred as 'pitches per inch' (ppi) then:

$$\text{PPI} = \frac{1}{\text{Lead}} \times \text{Number of starts}$$

1b.   Or, more simply, if the lead is expressed as tpi, and the pitch is required as ppi, then:

$$\text{PPI} = \text{TPI of lead} \times \text{number of starts}$$

2.   When the pitch is adopted as the base reference, then the LEAD will vary according to the number of starts:

$$\text{Lead (linear)} = \text{Pitch} \times \text{Number of starts}$$

2a.   If the lead is preferred in terms of tpi, then:

$$\text{TPI of lead} = \frac{1}{\text{Pitch} \times \text{number of starts}}$$

2b.   More simply, if we first resolve the pitch into pitches per inch: ppi, then:

$$\text{TPI of lead} = \frac{\text{PPI}}{\text{Number of starts}}$$

METRIC MULTIPLE START THREADS

Because with metric working there is no call for designating multiple start threads other than by pitch and lead, and because metric pitch and lead figures seldom involve a string of almost meaningless decimal figures as with English working, there is less chance of confusion, hence:

$$\text{LEAD} = \text{Pitch} \times \text{Number of starts}$$

$$\text{PITCH} = \frac{\text{Lead}}{\text{Starts}}$$

A further simplification lies in the fact that, unlike English tpi, many standard metric pitches are exact multiples of each other, with the result that leads are more often whole or integral numbers, free from recurring decimals.

MULTIPLE-START THREADING

Semi-Automatic and Automatic Start-Indexing

Let us carry out some simple practical experiments that can be performed on any small industrial-type lathe with either an English or metric leadscrew. The experiments will greatly assist in understanding all further explanations.

102

Experiment No.1 . 4 tpi leadscrew
Chuck a $3\frac{1}{2}$ inch length of 1 in. dia. free-cutting bright steel with a projection of about $1\frac{1}{2}$ in. from the chuck jaws. Face and chamfer the end. Skim a length of about $1\frac{3}{8}$ in. to a bright concentric finish. Set a Vee-type threading tool with a fairly sharp point, and form a runout groove at X (Fig. 14 Section 5) about $1\frac{1}{4}$ in. from the chamfered end, and of about 30 thou. in. depth and $\frac{3}{32}$ in. wide at the base. (This is only a demonstration run, and a full thread need not be subsequently cut, so we can take liberties with the 'threading' tool).

Set the selective gearbox to cut a thread of 6 tpi. Set the lathe spindle to run very slowly: about 80 rpm. Advance the threading tool to the component surface until it just scratches. Set the cross-feed dial to zero. Traverse the tool clear of the component to an approximate Y position. Advance the cross slide by 10 thou. in. Engage the half-nuts. Start the lathe. The tool will trace a 6 tpi helix. Be ready to stop the lathe *before* disengaging the half-nuts and when the tool reaches about the middle of the runout groove.

Set a left-hand buffer-type stop-indicator (as in Fig. 16 section 5) to abut the carriage side. (Perhaps a stop-indicator is not strictly necessary for such slow 'threading' speeds on external work whereon the approach of the tool to the runout groove can be watched, although on the other hand, watching helices in motion can give the illusion that they are moving axially past an already stationary tool, then, by the time it is realised that the tool is in fact still traversing, it may be too late to prevent an overrun.).

Disengage the half-nuts, rack traverse the carriage to the Y position with X – Y = 6 leadscrew threads: $1\frac{1}{2}$ in. Re-engage the half-nuts. Set a right-hand carriage dead stop C with a small gap G – the

latter for reasons already explained. Advance the threading tool to its original 10 thou. in. depthing. Start the lathe. Note that the tool follows the original helix. Be ever ready, however, to stop the lathe at X before disengaging the half-nuts. Stop the lathe at X. Disengage the half-nuts, and rack traverse up the R.H. carriage stop – or probably just short of it. Re-engage the half-nuts and reset the tool to the same 10 thou. in. depthing. Take a third cutting pass. Note to your satisfaction that every time these operations are repeated, the tool follows the original first helix trace.

Pick-up is held in this way because X – Y holds exactly 9 component thread turns to 6 leadscrew thread turns. The fact that some component thread turns are not cut in the Z gap makes no difference.

Now let us see what happens if instead of operating with an X – Y setting of 6 leadscrew threads ($1\frac{1}{2}$ in.) we move the stop C back $\frac{1}{4}$ inch, making the X – Y resetting distance 7 leadscrew threads, i.e. $1\frac{3}{4}$ in., thus deliberately misplacing pick-up.

Free the stop C, disengage the half-nuts, traverse the carriage one leadscrew thread to the right ($\frac{1}{4}$ in.), re-engage the half-nuts, reset and lock stop C, with a small gap G. Set the threading tool to its original 10 thou. in. depthing. Start the lathe and watch the component. You will find that the tool is tracing a second helix exactly midway between the first. Stop the lathe at X. Disengage the half-nuts. Rack traverse back to Y, re-engage the half-nuts and take another cutting pass at the 10 thou. in. depthing. Although you will not now be able to see what is happening, the tool will in fact re-trace the first helix. Repeat the resetting procedure, and the tool will re-trace the second helix, repeat again, and the tool will re-trace the first helix. In fact, if you continue thus, increasing the tool depthing at every *second* cutting pass you will ultimately cut a 2-start thread of 6 tpi *lead* (0.1666 . . . inch *lead*) and 0.08333 . . . in. *pitch:* the pitch having every appearance of a thread of 12 tpi — and you will have semi-automatically 'indexed' the two starts. I say 'semi-automatically' because the lathe had to be stopped at the termination of every cutting pass, and could not (indeed must not) be started again until the carriage is repositioned at the Y position with the half-nuts re-engaged.

START INDEXING THEORY

Exactly how does the lathe index the two starts in strict turn?

For a thread of 6 tpi from a leadscrew of 4 tpi, the basic gearing ratio is 2 to 3, i.e. each two leadscrew threads hold exactly three component thread turns. Accordingly, one leadscrew thread 'holds' exactly $3/2 = 1\frac{1}{2}$ component thread turns, therefore when we altered the X − Y setting distance to 7 leadscrew threads, the number of component thread turns 'embraced' was 7 LST × $1\frac{1}{2} = 10\frac{1}{2}$. Hence *every* cutting pass (after the first of any series, to be pedantic) is half a component thread turn out of phase with the *previous* cutting pass. The first cutting pass will be in phase (it cannot be otherwise). The second cutting pass will be $\frac{1}{2}$ turn out of phase with the first, thus a second helix is formed. The third cutting pass will again be half a turn out of phase with the second pass, so it must be in phase with the first pass: as you found.

Experiment No. 2. 4 tpi leadscrew.

Machine away the helices formed by experiment No. 1. Set the lathe to cut a thread of 7 tpi at 80 rpm.

Set the threading tool to the same X runout position as for No. 1 Exp. Cut a 7

tpi helix, 10 thou. in. deep. Stop the lathe at X. Traverse back until X − Y = 8 leadscrew threads (2 inches). Re-engage the half-nuts. Set stop C. Take a series of threading passes all at the 10 thou. in. depth. Note that the tool always follows the same helix. Pick-up for the single helix is held because 8 leadscrew threads span exactly 14 component thread turns (including the 'missing' turns in the Z gap).

After your last resetting at Y, deduct 2 leadscrew threads, making X − Y = 6 leadscrew threads. The basic ratio for a thread of 7 tpi from a leadscrew of 4 tpi is 4 to 7. Hence one leadscrew thread 'holds' 7/4 = $1\frac{3}{4}$ component thread turns, and 6 leadscrew threads therefore hold 6 × $1\frac{3}{4}$ = $10\frac{1}{2}$ component thread turns. Accordingly, if cutting passes are now made at the X − Y = $1\frac{1}{2}$ in. setting you will ultimately cut a 2-start thread of 7 tpi LEAD (0.1428571 in. LEAD) and 0.0714285 in. PITCH (which will look like 14 tpi.).

After re-tracing the two helices at a 10 thou. depthing a number of times to satisfy yourself that one or the other is always followed, then after your last Y resetting, retract stop C, move the Y position to the right through *three* leadscrew threads, making the total X − Y = 9 leadscrew threads, = $2\frac{1}{4}$ inches.

With the half-nuts engaged at the new setting and the same 10 thou. depthing, start the lathe and watch the component. You will see a third helix formed amongst the first two. Stop at X, disengage the half-nuts, traverse back to Y, re-engage the half-nuts, and take another cutting pass. This pass will re-trace one of the first two helices already cut, but on again resetting at Y and taking another cutting pass, a fourth helix will be formed, thus filling all gaps. Hence, if you continue thus, increasing tool depth only at every FOURTH cutting pass, you will ultimately form a 4-start thread of 7 tpi lead, and

0.0357142 in. PITCH (which will of course have every appearance of 28 tpi, except that the helix angle ('slope' of thread) will be more pronounced).

The reason for the formation of the four starts (or four separate helices) is that, as we have seen, with a ratio of 4/7, each leadscrew thread holds $1\frac{3}{4}$ component thread turns, therefore at the 9 leadscrew $X - Y$ setting we have embraced $9 \times 1\frac{3}{4} = 15\frac{3}{4}$ component thread turns, thus the Y position is permanently $\frac{1}{4}$ component thread turn out of phase with the leadscrew. You could also say perfectly legitimately that the Y position is permanently $\frac{3}{4}$ component thread turn out of phase. If we commence with a plain unthreaded workpiece:

The first pass will cut No.1 helix Phase '1'
The 2nd ,, No. 2 helix $\frac{1}{4}$ or $\frac{3}{4}$
The 3rd ,, No. 3 helix $\frac{1}{2}$
The 4th ,, No. 4 helix $\frac{3}{4}$ or $\frac{1}{4}$

Then at the fifth pass we are already $4 \times \frac{1}{4} = 1$ whole turn out of phase, so the fifth pass will recut the first helix, the sixth pass will cut the second helix, and so on.

Experiment No.3. Metric Leadscrew. 6.0 mm pitch.

Prepare an 88 mm length of 25 mm dia. (or 1 in. dia.) free cutting mild steel as for No. 1 exp., but let the length to be threaded equal 30 mm (about $1\frac{1}{8}$ in.) from the chamfered end to the runout groove.

Set the lathe to cut a thread of 4.0 mm lead at about 80 rpm. Take a threading pass of about 0.25 mm depth. Stop the lathe at the X runout position. Disengage the half-nuts and rack traverse to position Y with $X - Y = 6$ leadscrew threads $= 36$ mm thus offering a Z starting clearance of 6.0 mm. Set the right hand stop C, with a small gap G. Take a number of threading passes all at the same 0.25 mm depth and satisfy yourself that pick-up is assured.

Release stop C and move the carriage to the right through one leadscrew thread, making $X - Y = 42$ mm, re-engage the half-nuts and reset stop C. Adjust tool depthing to the same 0.25 mm and take a threading pass. You will see another helix formed mid-way between the first.

Minimum pick-up for a lead of 4.0 mm with a leadscrew of 6.0 mm pitch is 2 leadscrew threads to 3 component thread turns, hence one leadscrew thread holds $3/2 = 1\frac{1}{2}$ component thread turns, accordingly, with $X - Y$ set at the seventh leadscrew thread the relationship is altered to 7 leadscrew threads $\times 1\frac{1}{2} = 10\frac{1}{2}$ component thread turns, consequently the $X - Y$ setting is one half a component thread turn out of phase, and if this setting is repeatedly followed, then you will cut a 2-start thread of 4.0 mm LEAD and 2.0 mm PITCH.

Experiment No. 4. Metric leadscrew 6.0 mm pitch. To index 3-starts.

Machine away the threads formed at Exp. No. 3. Set the lathe to cut a thread of 4.5 mm lead at about 80 rpm. Set $X - Y = 7$ leadscrew threads $= 42$ mm. Set stop C etc. Take a series of threading passes at 0.25 mm depth, always stopping the lathe at the X runout position before disengaging the half-nuts.

You will see that at the first cutting pass, one helix will be formed. At the second cutting pass, a second helix will be formed, and at the third cutting pass, a third helix will be formed. What you will not be able to see is that at the fourth cutting pass, the first helix will be re-traced. However, if you increase tool depthing only after every set of three cutting passes you will ultimately cut a 3-start thread of 4.5 mm *lead* and 1.5 mm *pitch*.

Minimum pick-up for a single 4.5 mm lead is 3 leadscrew threads to 4 component thread turns, so one leadscrew thread holds $4/3 = 1\frac{1}{3}$ component thread turns and 7 leadscrew threads therefore hold $7 \times 1\frac{1}{3} = 9\frac{1}{3}$ component thread turns, so we are permanently $\frac{1}{3}$ component thread turn out of phase, with results with which you are now familiar.

Experiment No. 5. 6.0 mm pitch leadscrew. To index 5-starts.

Using the same workpiece as for Exp. No. 4, clean off the threads to a smooth surface. Set the lathe to cut a thread of 5.0 mm lead, and set the spindle to run at about 60 rpm. Take a threading pass at a depth of about 0.25 mm. Stop the lathe at X, disengage the half-nuts and run the carriage to the right through 6 leadscrew threads = 36 mm. Re-engage the half-nuts and set stop C. Now take four more threading passes, carefully following the 'stop-and-reset' routine and you will form five separate helices, and, of course, if you persist at this X − Y setting, increasing tool depth on completion of every fifth cutting pass, you will fully form a 5-start, 5.0 mm *lead*, 1.0 mm *pitch* thread.

Basic gearing for a lead of 5.0 mm with a leadscrew of 6.0 mm pitch is 5/6, hence each leadscrew thread holds $6/5 = 1\frac{1}{5}$ component thread turns, accordingly 6 leadscrew threads hold $6 \times 1\frac{1}{5} = 7\frac{1}{5}$ component thread turns. The X − Y setting is therefore permanently $\frac{1}{5}$ component thread turn out of phase.

FULLY AUTOMATIC START INDEXING

Had the foregoing experiments been carried out on lathes with single-tooth dog-clutch control for the leadscrew drive, it would not have been necessary to stop the lathe spindle at every X runout position. The clutch would automatically arrest carriage traverse by stopping the leadscrew, thus relieving the operator of all anxiety over stopping at the correct position, hence, after making an initial X − Y setting, the operator has only to count the cutting passes, according to the number of starts, before making a depthing increment, and the operation thereafter is no different from the cutting of an ordinary single-lead thread. This process is entirely free from any urgency. The number of revolutions made by a workpiece after clutch disengagement and before re-engagement for the next cutting pass has no effect whatsoever on the sequence or order in which individual helices come into phase for cutting. Thus, for example, if, after resetting at Y, the second of three helices is due to be traversed, it will be traversed, even if in the meantime the workpiece has been rotating throughout the whole of a lunch-break. It is the dead' leadscrew that will have been waiting 'out of phase', the disengaged dog-clutch will merely be waiting for a 'whole component thread turn pick-up', accordingly it is of no consequence at which of any series of component turns the clutch is re-engaged to set the leadscrew in motion. However, at the risk of being repetitive I will say again that when using this method, all cutting passes must be made in full, right up to the X position, or up to automatic clutch disengagement, if the lathe has a clutch, otherwise pick-up will be lost. One sometimes forgets to re-engage the half-nuts at the Y position, too, and then again, if the leadscrew is set in motion, pick-up will be lost. It has to be 'found' again by taking up all gearing slack and making trial half-nut engagements along the length of the workpiece until the tool can be seen to coincide with one of the thread grooves, after which traverse must again be completed up to the X runout position. Nut threads are very difficult to re-synchronise, so extra care should be taken when dealing with these.

By use of the methods outlined, the cutting of any given multiple-start thread with several shallow helices takes virtually no longer than the cutting of a much deeper single thread of the same lead, and with the higher threading speeds offered by clutch control and use of an independently retractable toolholder, multiple-start threads can be produced very rapidly.

But even on lathes without dog-clutch control for the leadscrew drive, the 'stop-and-reset' method has distinct advantages over all other means for start indexing (except perhaps one, which uses multiple tools spaced at an exact pitch distance apart). Ordinarily one is advised to complete one start fully (i.e. one helix) before commencing on the next. While this may not be totally unsatisfactory for cutting a few multiple start *screws* – where the individual helices can be checked for size by 'wiring' (q.v.) it presents difficulties and uncertainties when internally multiple-start threading simply because a multi-start nut thread cannot be checked for size until all starts have been completed, or are thought to have been completed. And then, if a screw gauge will not enter, which of the starts is undersize? There is no way of telling, and all starts would have to be laboriously individually re-indexed and shaved until the gauge did fit: obviously an extremely time-consuming task, with risk of leaving one or more of the starts oversize. On the other hand, using the 'stop-and-reset' method, tool wear is evenly distributed over all starts, so individual helices are more likely to be of identical size to close limits, and if a gauge does not fit, one has only to take the appropriate number of cutting passes, each set of passes with a small depthing increment, until a gauge does fit: any question of re-indexing does not come into the operation.

IS AUTO-START INDEXING ALWAYS POSSIBLE?

We have seen that to automatically index the starts of multiple-start threads we have to deliberately misplace pick-up for a single lead. This precludes all possibility of auto-start indexing any thread the lead of which is the same as or an exact multiple of the leadscrew pitch, because for such threads the minimum leadscrew threads figure is 1 (one) which means that at whatever position the half-nuts are re-engaged on a leadscrew, only a single lead will be cut on a workpiece. Thus, for example, a lead of 1/8 in. (8 tpi) when cut from a leadscrew of 8 or 4 tpi cannot be auto indexed into any starts except one, because pick-up from an 8 tpi leadscrew is 1 LST to 1 WT, and from a leadscrew of 4 tpi, pick-up is 1 LST to 2 WT. There are however two methods for overcoming these limitations, and these will shortly be described. In future, to avoid wearisome repetition of 'Auto-Start' & 'Semi Auto-Start' both will be referred to as 'Auto-Start' Indexing.

FEASIBILITY TEST

The rules for ascertaining whether or not auto-start indexing is possible are simple:

A. If the LST figure (numerator) in any basic gear ratio is exactly divisible by the number of starts to be indexed, then auto-start indexing is possible.

B. The quotient from test *A* (when division is possible) gives the number of leadscrew threads which should be added to or deducted from the total number of leadscrew threads in pick-up for a single lead.

For example, with a leadscrew of 8 tpi, threads of 5, 7, 9, 11 etc. tpi can be auto indexed into 2, 4 or 8 starts, simply because 8 (the LST figure in the basic ratios 8/5, 8/7, 8/9, 8/11) is divisible by 2,

4 and 8. The X-Y settings would be:

For 2-starts, $8/2 = 4$, set 8 plus or minus 4.

For 4-starts, $8/4 = 2$, set 8 plus or minus 2.

For 8-starts, $8/8 = 1$, 8 plus or minus 1.

NOTE: according to the length to be threaded, complete the total X – Y units that would give a single lead, then add or deduct the leadscrew thread displacement figure.

For example: required a 4 in. length of 5 tpi, 4-start 0.05 in. pitch thread from a leadscrew of 8 tpi. 4 in. X 8 = 32 LST = X – Y pick-up for a single lead. $8/4 = 2$, add two leadscrew threads, making total X – Y = 34 LST = $4\frac{1}{4}$ in.

Example 2. A leadscrew of 6.0 mm pitch is to be set to cut a 100mm length (about 4 in.) of 4.5 mm lead, 3-start, 1.5 mm pitch. Find the X – Y setting.

The basic gear ratio will be $4.5/6 = 3/4$, i.e. 3 LST to 4 WT. Pick-up for a single lead therefore falls into groups of 3 LST, and 6 groups $= 6 \times 18 = 108$ mm. Misplacement for three starts $= LST/3 = 1$. Therefore deduct one leadscrew thread from the six groups, and X – Y = five groups plus 2 LST = 17 LST = 102 mm.

If a 2.0 mm starting clearance is insufficient, then we can add one leadscrew thread to the six groups of three, making X – Y = 19 LST = 114 mm.

As a matter of fact when the LST figure is a prime as in this example (prime 3), three starts can be indexed by misplacing plus or minus one or *two* leadscrew threads. One LST holds $4/3 = 1$ & 1/3 WT, and 2 LST hold 1 & 1/3 × 2 = 2 & 2/3 and as far as we are concerned, being 2/3 out of phase has exactly the same effect as being 1/3 out of phase. Similar reasoning will show that when the LST figure in basic ratio is prime 5, a LST displacement of 1, 2, 3 or 4 leadscrew threads from

pick-up for a single lead will auto index 5-starts.

Later we will encounter LST figures composed of two prime elements, such as 33 = prime 3 × prime 11. As 33 is divisible by 3 = 11, a LST misplacement of plus or minus 11 LST would index three starts, but you will find that we can also displace plus or minus 22 LST for three starts. If a basic gear ratio is for example 33/35 then 1 LST holds 35/33 WT (leave in improper fractional form) and 11 LST therefore hold $35/33 \times 11 = 11$ & 2/3 WT, and 22 LST hold $35/33 \times 22 = 23$ & 1/3 WT.

It is worth noting that in all these calculations we are hardly interested in whether a leadscrew is of English or metric (or any odd) standards: we deal almost entirely with the relationship leadscrew-thread-turns-to-component-thread-turns for any particular gearing ratio. Thereafter, if for example, an X – Y setting has to include 6 leadscrew threads it is useful to know that if the pitch is of 6.0 mm, six such threads make X – Y = 36 mm and if the pitch is 0.25 in., then six leadscrew threads make X – Y = 1.5 in.

Perhaps I should add that although all the foregoing may appear somewhat complicated when judged by the amount of writing required to describe the process, in practice, as soon as minimum pick-up from basic gear ratios for any single lead has been mastered, then displacement and subsequent X – Y setting requires little more than a quick mental calculation, or at most, a few figures on a scrap of paper.

CHECKING AN X – Y SETTING

That any proposed setting will automatically index the required number of starts (assuming the gearing is correctly set) may be checked by the following means which is applicable to English leadscrews, metric leadscrews, and to leadscrews of non-standard metric or English pitch (q.v.).

Invert the basic gear ratio to Driven/Driver and multiply by the number of leadscrew threads held in the proposed X – Y setting. Do not resolve improper fractions. If the X – Y setting is correct, then the denominator in the quotient will equal the number of starts required.

Example 1. A 4 tpi leadscrew is set to cut a 3 tpi lead, 2-start 0.1666 in. pitch over a length of $1\frac{1}{4}$ in. The basic gear ratio is 4/3. The X – Y setting should be 6 LST = $1\frac{1}{2}$ in. Is this correct?

Checking:
$$\frac{\text{Driven } 3}{\text{Driver } 4} \times 6 \text{ LST X} - \text{Y} = \frac{9}{2}$$

Denominator 2 shows that 2 starts would be produced.

Example 2. A leadscrew of 6.0 mm pitch is geared 18/17 to cut nine starts, 4 tpi lead, 0.02777 in. pitch over a length of 90 mm. The X – Y setting was calculated to = 16 LST = 96 mm.

Checking:
$$\frac{\text{Driven } 17}{\text{Driver } 18} \times 16 \text{ LST X} - \text{Y} = \frac{136}{9}$$

Denominator 9 shows that 9 starts would be produced.

Example 3. A special leadscrew of 7 1/3 tpi is geared 4/3 to cut two starts 0.0909 in. pitch on a lead of $5\frac{1}{2}$ tpi over a thread length of 2 in. X – Y is set at 18 LST (about 2.45 in.)

Checking:
$$\frac{\text{Driven } 3}{\text{Driver } 4} \times 18 = \frac{27}{2}$$

Denominator 2 shows that the proposed setting will index 2 starts.

Example 4. A leadscrew of 3.5 mm pitch is geared 39/43 to cut a lead of 1/8 in. (8 tpi), 3-starts, 1/24 in. pitch. The length to be threaded is $1\frac{5}{8}$ in., and the proposed X – Y setting is 13 LST = 45.5 mm. (The gearing happens to be a basic ratio, 43 being a prime.)

Checking:
$$\frac{\text{Driven } 43}{\text{Driver } 39} \times 13 \text{ LST} = \frac{43}{3}$$

Denominator 3 confirms that three starts would be indexed.

AUTO START INDEXING – ALTERNATIVES

1. Use of leadscrew of opposite language to thread to be auto-indexed.
2. Use of leadscrews of special lead.

(1) When a thread in the same language as that of a lathe leadscrew cannot be auto-indexed into starts because the LST figure in minimum pick-up is not exactly divisible by the number of starts required, it is worth looking into the possibility of cutting the thread from a leadscrew of opposite language wherein the necessary gearing (generally derived from an approximate translation ratio) may show an LST figure that is divisible by the number of starts to be indexed.

As we have seen, a lead of 0.25 in. (4 tpi) cannot be indexed into any starts except one, when cut from a leadscrew of 4 tpi, but if a lathe with a leadscrew of 6.0 mm pitch is geared in the basic ratio 18/17 (See Table T11A&B Section 5), then, as numerator 18 is divisible by 2, 3, 6, 9 and 18, all those starts could be auto indexed. (The lead error with the 18/17 ratio is plus 0.0001157 in. assuming a leadscrew of perfect lead).

For a five start thread of the same 0.25 in. lead we could use the 55/52 ratio (pitch error minus 0.0001515 in.) And, of course, as 55 is divisible by 11, the ratio could be used to auto index 11 starts if required.

If we have to index the 0.25 in. lead into four starts, then we require an approximate ratio with a numerator divisible by 4, so we could use either 52/49 (error plus 0.00068 in. on pitch), or 88/83 (error plus 0.00045 in. on pitch). Also, as 52 is divisible by 13, then 13 starts could be auto indexed.

■ Similarly, metric leads can be start indexed by making use of the approximate ratio for metric threads cut from a leadscrew of 4 tpi. (Table T10A&B Section 5).

For example, a lead of 6.0 mm cannot be auto indexed into any starts except one when cut from a leadscrew of 6.0 mm pitch, but if the required 6.0 mm lead is geared from a leadscrew of 4 tpi in the ratio 52/55, then, as 52 is divisible by 2, 4 and 13, those starts could be indexed on a 6.0 mm lead (lead error plus 0.003636 mm (0.0001431 in.)). And if we want to index the 6.0 mm lead into three starts we have a choice of two numerators divisible by 3: 33/35 (error minus 0.0128 mm on lead) or 69/73 (error plus 0.00205 mm on lead). Finally, a 6.0 mm lead can be indexed into five or seven starts by use of the 35/37 ratio (lead error plus 0.00675 mm) In short, the selection of a suitable approx ratio is merely a matter of scanning the numerators in the Tables T10&T11 and making the divisibility test.

INCONVENIENCES

However, with this approximation approach, inconveniences can arise from the length of the Z gap (starting clearance) when the length of thread to be cut and start indexed is incompatible with a necessary X – Y setting.

For example if a 1 in. length of 6.0 mm lead, 3-start, 2.0 mm pitch is required and the 33/35 ratio is used in conjunction with a leadscrew of 4 tpi, the minimum X – Y setting is 33/3 = 11 leadscrew threads = 2.75 in., thus leaving an approximate 1.75 in. Z gap to be re-traversed after every resetting at Y. And if we choose the 69/73 ratio with its theoretically smaller error, 69/3 = 23 LST = $5\frac{3}{4}$ in. X – Y minimum, which would leave a gap of about $4\frac{3}{4}$ in. to be repeatedly re-traversed.

Nevertheless, in view of the fact that start indexing would be automatic, or semi-automatic, and cutting the threads merely involves counting sets of cutting passes before adding depthing increments, then when only one or a few such threads are required, the approximate ratio approach has much to recommend it, even with fairly lengthy Z gaps and a necessity for ordering any special gears for setting on an 'outside' quadrant.

MULTI-START THREADS. QUANTITY PRODUCTION. SPECIAL LEADSCREWS

(2) In those instances where a significant number of multi-start threads of identical lead and pitch have to be cut and any of the means already outlined cannot be used for any reason, then a great saving in time can be effected by making a leadscrew and half-nuts of special lead, the leadscrew, of course, having just sufficient thread to suit the component thread length, with the remainder as plain shaft.

An example of a special leadscrew arose when the writer was called upon to thread 150-off, 2-start, $5\frac{1}{2}$ tpi lead, 0.0909 in. pitch screws *and* blind bore nuts.

Although a lead of $5\frac{1}{2}$ tpi can be auto-indexed into two starts from a leadscrew of 8 tpi: basic ratio 16/11, the minimum X – Y setting would have been 8 LST = 1 in. which would have fallen at 16 plus 8 = 24 LST = 3 in. which would have left a wastefully long Z gap to be re-traversed about three thousand times. The idea of indexing the starts by orthodox means (q.v.) was of course dismissed as being totally uncompetitive, especially as repeat orders were anticipated. Therefore, in the event, a special leadscrew of 7 1/3 tpi was made, with a threaded length of about 5 in., which brings us to the subject of special leadscrew design.

SPECIAL LEADSCREW DESIGN

As has been explained, if the number of leadscrew threads in basic gear ratio for a single lead is exactly divisible by the number of starts to be auto-indexed, then auto-indexing is possible. From this it follows that if we predetermine the pick-up figures (minimum LST numerator figures) to suit our special requirements, then calculate what metric pitch, or threads/inch a leadscrew will assume to hold those figures, we can design individual leadscrews that will automatically index any lead into any number of starts with economical Z gap starting clearances regardless of the lengths of the component threads.

SPECIAL LEADSCREWS. ENGLISH WORKING

Because of the awkwardness that can be introduced by recurring decimals when English special leadscrews are designated by lead, it is preferable to express lead in terms of threads/inch as an integral number, or as a mixed number fraction. The formula reads:

TPI of special
leadscrew

$$= \frac{\text{TPI of lead} \times \text{LST No. chosen}}{\text{WT number chosen}}$$

where TPI of lead is the lead tpi of the workpiece, LST is the numerator figure

chosen for basic ratio, and WT is the component thread turns denominator figure for basic ratio. The rules are as follows:

Choose an integral WT number that will not cancel or reduce with an integral number of leadscrew threads (the LST figure), the latter being given a value the same as, or an exact multiple of the number of starts required.

In choosing the LST and WT figures for special leadscrews, select the lowest possible LST figure so as to minimise Z tool-starting clearances, and at the same time arrange for a LST to WT ratio that will offer a leadscrew pitch near to the standard for the lathe, otherwise a special leadscrew may prove to be undesirably coarse or fine.

Example 1. Find a suitable leadscrew tpi to auto index a $5\frac{1}{2}$ tpi lead into two starts, 0.0909 in. pitch.

Let LST $= 4$ and WT $= 3$

TPI of special leadscrew $= \dfrac{5\frac{1}{2} \times 4}{3} = 7\frac{1}{3}$

In this example, 4 was chosen for the LST figure because it is divisible by 2, the number of starts to be auto indexed, and a WT figure of 3 was selected because it will not cancel with 4.

The special leadscrew (and half-nuts) was cut from a leadscrew of 8 tpi by basic gearing 12/11. then, for the necessary $5\frac{1}{2}$ tpi lead, the new leadscrew was geared in the ratio 4/3. Now, of course, the basic ratio 4/3 is that originally chosen, so with the special leadscrew we have 4 LST to every 3 component thread turns, hence a displacement of plus or minus 2 LST on any X – Y setting for a single lead (groups of 4 LST) indexed the two starts on the $5\frac{1}{2}$ tpi component lead, and even under the most unfavourable conditions, any Z starting clearance gap could never exceed 4 LST $= 0.5454$ in.

We may also note that as the LST figure of 4 is divisible by 4, the special leadscrew could be used to index the $5\frac{1}{2}$ tpi lead into four starts by a LST displacement of plus or minus 1 LST on pick-up for a single lead.

Example 2. Find a suitable leadscrew tpi to auto index 2-starts, 1/8 in. lead (8 tpi) 1/16 in. pitch.

Let LST = 4, and WT = 5

$$\text{TPI of special leadscrew} = \frac{8 \times 4}{5} = 6\frac{2}{5}$$

This leadscrew would therefore auto index the two starts by an LST displacement of 2 on pick-up for a single lead. A displacement of 1 or 3 leadscrew threads would also auto-index 4-starts, 1/32 in. pitch.

However, had we chosen a WT number to contrast with an LST figure of 6, the resulting leadscrew would automatically index an 8 tpi lead into 2, 3 or 6 starts:

Let LST = 6 and WT = 7

$$\text{TPI of special leadscrew} = \frac{8 \times 6}{7} = 6\frac{6}{7}$$

Example 3. Required a special leadscrew to index 5-starts on a lead of 3 tpi. (0.0666 in. pitch).

Let LST = 5 (divisible by 5 for indexing 5 starts)

Let WT = 4

$$\text{TPI of special leadscrew} = \frac{3 \times 5}{4} = 3\frac{3}{4}$$

In all examples, the basic gear ratio to cut the required lead tpi from the special leadscrew will be that chosen when designing the leadscrew: LST/WT. Thus in example 3, when the $3\frac{3}{4}$ tpi leadscrew is ready it should be geared 5 to 4 to cut the 3 tpi lead.

(Proof: driven/driver x tpi of leadscrew = $4/5 \times 3\frac{3}{4} = 4/5 \times 15/4 = 3$.).

SPECIAL LEADSCREWS – METRIC.

The pitch of a special leadscrew for auto-start indexing leads specified to metric standards can be found from:

Special leadscrew Pitch (mm) =

$$\frac{\text{Lead of screw to be cut} \times \text{WT chosen}}{\text{LST figure chosen}}$$

where the lead of the thread to be cut is expressed in mm. The rules are as follows:

Choose an integral work thread turns number (WT) that will not cancel or reduce with an integral number of leadscrew threads (LST), the latter being given a value the same as or an exact multiple of the number of starts to be indexed.

Example 1. Required a suitable leadscrew pitch to auto-index 5-starts, 3.5 mm lead, 0.7 mm pitch.

As we require five starts it will be reasonable to select 5 for the LST figure, and, say, 4 for the WT figure, so that whatever pitch the special leadscrew assumes, it will offer a minimum pick-up of five LST to 4 component thread turns: WT

$$\text{Pitch of special leadscrew} = \frac{3.5 \times 4}{5} = 2.8 \text{ mm}$$

In this example, had we chosen 7 WT to 5 LST the pitch of the special leadscrew would be:

$$\frac{3.5 \times 7}{5} = 4.9 \text{ mm}$$

We could also have chosen 3 WT to 5 LST which would offer a special leadscrew pitch of:

$$\frac{3.5 \times 3}{5} = 2.1 \text{ mm}$$

which could prove inconveniently fine.

Example 2. Required a special leadscrew to auto index a lead of 6.0 mm into 3-starts, 2.0 mm pitch. We have seen how this indexing can be carried out by using an English leadscrew of 4 tpi geared, e.g. in the ratio 33/35, but we also noted that for some component thread lengths the *Z* gap could be wastefully long for quantity production. Accordingly, for a special leadscrew offering economical starting clearances for threading components of any length:

Let LST = 6 (divisible by 3, for 3-starts)
Let WT = 5

Then pitch of special leadscrew $= \dfrac{6 \times 5}{6} = 5.0$ mm

and of course, a leadscrew of 5.0 mm pitch geared in the ratio 6 to 5 will cut a thread of 6.0 mm lead, and a displacement of plus or minus 2 (or plus or minus 4) LST on pick-up for a single lead will index three starts.

(A displacement of plus or minus 3 LST would index 2-starts on the 6.0 mm lead, and a displacement of plus or minus 1 LST would index six starts.)

Example 3. Required a suitable leadscrew pitch to auto-index 3-starts, 3.0 mm lead, 1.0 mm pitch.

The leadscrew of 5.0 mm pitch (from example No. 2) would serve when geared in the basic ratio 3 to 5 showing 3 LST to each 5 component thread turns. Accordingly a displacement of plus or minus 1 or 2 leadscrew threads on pick-up for a single lead would index the three starts. Otherwise:

Let LST = 3 and WT = 4, then:

Pitch of special leadscrew $= \dfrac{3 \times 4}{3} = 4.0$ mm

However, a 4.0 mm pitch leadscrew with 3 LST to 4 WT would only index a 3.0 mm component lead into 3-starts or a single lead, whereas had we chosen an LST figure of 6 any resulting leadscrew would auto index 2, 3 and 6 starts, so let us try LST = 6 and WT = 7, the special leadscrew pitch will then be:

$$\frac{3 \times 7}{6} = 3.5 \text{ mm}$$

You will notice that in this example (3) we have a choice of three leadscrew pitches for indexing a 3.0 mm lead into 3-starts: 5.0 mm, 4.0 mm, and 3.5 mm. This was one of the circumstances that led the writer to question the standardisation of leadscrews of 3.0 mm pitch for lathes of the instrument or model makers' type and size, and after considerable trial calculations the conclusion was reached that for general purpose threading, and for special purposes such as auto-start indexing, a standard leadscrew of 3.5 mm pitch offered the widest scope.

It so happened that shortly after this 3.5 mm pitch leadscrew was ready for use, an order was received for 100-off 5.0 mm lead, 2-start, 2.5 mm pitch screws and nuts. Basic gearing for a lead of 5.0 mm from a leadscrew of 3.5 mm pitch is 10 driver, 7 driven, hence the LST figure of 10 was divisible by 2, and the two starts could be auto-indexed by a displacement of plus or minus 5 leadscrew threads, so any Z gap could not exceed 35 mm or about 1.3 inches.

Had the leadscrew been of 3.0 mm pitch, the basic gear ratio would have been 5/3, i.e. 5 LST to 3 WT, and as 5 is not divisible by 2, auto-start indexing would not have been possible.

STEP-UP QUADRANT GEARING

Although the use of step-up ratios whereby a leadscrew is caused to rotate at speeds in excess of the lathe spindle are not to be generally recommended, there is the point that when such ratios are used for cutting coarse leads associated with multiple-start threads, the loading on a lathe is not so severe for the reason that the individual helices are sized and depthed by pitch, and the greater the number of starts carried by any particular lead, so in proportion will the thread form be shallower and therefore more easily cut.

As already mentioned, however, step-up ratios cannot always be used in association with the selective threading gearboxes sometimes fitted to small lathes of the model makers' type.

MULTIPLE-START THREADING

Six alternative start-indexing methods.

1.	As already mentioned, use two or more threading tools spaced at exact pitch distance apart. For the finer pitches and small bore internal threading it may be necessary to have integral tools made up by a specialist firm. Components to be threaded must be free from shoulders, or must have ample runout clearance to allow tooling to run completely clear of the workpiece threads. Components must have sufficient rigidity to withstand the multiple cutting action. The multi-tool approach is therefore more readily applied to the larger diameters on materials of the brass type which are easy to thread to a good finish. Suitable for both screws and nuts.

2.	The starts may be indexed by means of a chuck that can be independently rotated relative to its backplate, the backplate being provided with a series of equi-spaced holes. Complete one start of any series before commencing the next. Suitable for screws, and, with reservations already explained (inability to check for size until all starts are completed) may be used for nuts.

3.	For work that can be driven between centres, use a driving plate with a series of equi-spaced holes for the driving pin. Complete one helix before commencing the next. Suitable only for screws.

4.	Arrange for the lathe spindle gear (first gear driver to a leadscrew gear train) to have a number of teeth divisible by the number of starts required. Complete one start before commencing the next. Index the starts by de-gearing, rotating the lathe spindle through the appropriate number of teeth on the first gear driver, and re-engaging the gears. Suitable for screws, and, with reservations, for nuts.

5.	Operate with the top-slide set parallel to the lathe bedways, and on completion of one start, advance the top slide through one pitch distance (by reading the top slide feed dial), then cut the next start. The top-slide may also be advanced by small amounts to ease the trailing cutting action of the threading tool, but any such advancements should be exactly repeated for all helices. This method requires a run out of sufficient width to allow for all necessary top-slide pitch advancements. Suitable for screws, and with reservations, for nuts.

NOTE: For methods No. 1 to 5, pick-up for lead may be held by any of the means already described.

6.	Start index by use of a leadscrew indicator (LSI). When pick-up can be displaced it is sometimes possible to use a leadscrew indicator to pick-up alternate starts by 'cutting in' (re-engaging the half-nuts) on 'wrong' LSI indications.

For example, for the $5\frac{1}{2}$ tpi lead, 2-start, 0.0909 in. pitch threads for which a special leadscrew was made (page 110) a leadscrew indicator could have been used in conjunction with the standard leadscrew of 8 tpi.

Pick-up for a single $5\frac{1}{2}$ tpi lead from a leadscrew of 8 tpi is 16 leadscrew threads (or 16 leadscrew revolutions) i.e. one whole turn of a 2-inch LSI. Consequently had one cutting pass been made at LSI reading 1 (one), then one helix would be traversed, then for traversing the next helix it would be necessary to engage the half-nuts at LSI reading 3, i.e. half a turn of the LSI dial = 8 leadscrew threads, or eight leadscrew revolutions, the displacement figure for two starts. Similarly, re-engaging the half-nuts at 'wrong' LSI readings 1, 2, 3 and 4 would index four starts on the work-piece because the LST displacement for 4-starts = 16/4 = 4 leadscrew threads, or four leadscrew revolutions = $\frac{1}{4}$ turn of a LSI.

The disadvantage with this method seems to lie in having to wait for a LSI to register the appropriate moments for half-nut re-engagement. Although a leadscrew will be making a few revolutions between the moment of half-nut disengagement at the X runout position and whilst rack traversing back to a Y starting position, there obviously can be no guarantee that a LSI will be approaching anywhere near or will not just pass the necessary reading. For example, during a test run, using a leadscrew indicator to index 2 (and 4) starts on a $5\frac{1}{2}$ tpi lead of about 1 in. length, the lathe spindle was rotating at about 80 rpm, the gear ratio to the lead-screw was 16/11, therefore the leadscrew was rotating at about 116 rpm = about 1.9 rev/sec., yet it was frequently necessary to wait whilst the leadscrew indicator made about 3/4 turn = 12 revolutions of the leadscrew: about 6 seconds. Therefore, assuming each helix required 12 cutting passes to complete, i.e. 24 passes for each 2-start thread, there is a potential $2\frac{1}{2}$ minutes per screw (or nut) completely wasted: about 4 hours per 100 off, or, with dextrous operation at 160 rpm — twice the test run speed — 2 hours could still be wasted. There is also the point that waiting for a LSI to register is little different from waiting for a tool to traverse a long adverse Z starting gap when using the stop-and-rack-reset method.

However, when pick-up permits, indexing starts from a LSI has the advantage that helices are alternately cut and progressively depthed and are therefore more likely to be completed to identical sizes: a feature of value for nut threading. It is also somewhat easier to disengage the half-nuts on completion of each threading pass than to stop the lathe spindle and leadscrew at the desired moment before disengaging the nuts.

We should note, however, that the thread to be start-indexed with a LSI must be of a lead for which a LSI could be used normally to show a single lead pick-up, and any necessary LST displacement figures must be exact sub-multiples of one whole turn of a LSI as were the 8 and 4 LST displacement figures for indexing the $5\frac{1}{2}$ tpi lead into two or four starts.

As an example for which a LSI is of no use, we may take a leadscrew of 8 tpi geared in the ratio 3 to 2, say 45—A—30, for cutting a lead of 3/16 inch. The ratio shows 3 LST to 2 WT minimum, hence, by the stop-and-rack-reset method, three starts, 1/16 in. pitch, could be auto indexed by re-engaging the half-nuts at any leadscrew position $X - Y$ *except* groups of 3 leadscrew threads. From this, it may be thought that if groups of 8 LST could be read from a LSI by engaging the half-nuts at, say, LSI readings 1 and 3 ($\frac{1}{2}$ turn of the LSI dial), three starts would be indexed.

A practical test clearly showed that three starts are obtained, but the snag was that the helix cut depended entirely upon the number of revolutions made by the workpiece and leadscrew during the non-cutting return passes. As a matter of fact, after engaging the half-nuts at LSI reading 1 (one) the first helix was cut, but for the next pass, after re-engaging the half-nuts at LSI reading 3, all that happened was that the first helix was re-traced. Ultimately, of course, by chance, all helices were formed, but not in any regular sequences, and indeed, the LSI was not performing any useful function at all: the same haphazard start indexing was achieved by re-engaging the half-nuts at any random moment and in any random position. This was possible because there were only three 'wrong' re-engagement positions, so it was not possible to re-trace other than any one of the three helices at any given cutting pass: at least it was not possible to spoil the threads.

Evidently, then, an English leadscrew indicator cannot be adapted to the systematic indexing of starts other than 2 and 4, and in somewhat limited instances.

METRIC LEADSCREW INDICATOR.

With a leadscrew of 6.0 mm pitch and a 20-tooth LSI worm wheel, theory indicates that the metric leadscrew indicator is of no use at all for systematically start indexing metric leads:

many standard leads pick up either at unity or at 5 LST intervals, and of course, 5 cannot be exactly halved.

SPECIAL TAPS

When any significant number of multi-start nut threads have to be produced, much time can be saved by making a tap for rapid finishing after lathe screwcutting to within about 90% of final size. I say 'making' rather than 'ordering', on the assumption that the reader will be cost-conscious. Some firms will not make less than six of any one size, and the cost could wipe out all profit from the first batch order, apart from time lost awaiting delivery of specials.

Single Point Lathe Threading Tools

The expression "single point" is used to distinguish the types of tools to be discussed from special thread form-tools and from the comb-type of chasers which can be obtained for the more specialised branches of lathe threading, and for threading at higher speeds in purpose built machines. When special thread form tools are felt necessary it is advisable to consult the makers so that the grade best suited to the work in hand may be obtained, and, of course, the makers' instructions for sharpening will be followed.

Lathe tools can be obtained in a number of materials and brief notes are now included to assist in identifying and choosing a suitable cutting medium.

HIGH SPEED STEEL (HSS)

Originally this was a high grade steel alloyed with tungsten in various proportions from about 10 to 20 percent. Nowadays, owing to the high cost of tungsten, various other alloying agents such as cobalt and vanadium are used extensively. For short run work on the more commonly used steels, brasses and bronze, high speed steel is very popular and should be found quite satisfactory.

STELLITE

This, I understand, is now obsolete, although pieces and tipped tools will often be found amongst odd lots of lathe tools. It is an alloy of the elements cobalt, chromium and tungsten. It is quite free from steel and has a hardness approaching that of tungsten carbide. It does not soften with heat and it can be ground with ordinary wheels suitable for high speed steel.

Although stellite had many important applications for general machining, in the experience of the writer it will not hold a fine point and was therefore found unsuitable for screwcutting the finer pitches in the harder steels. The alloy was expensive, and for this reason was often used in the form of tips mounted on ordinary steel shanks, although solid round and square sections could be obtained, usually in the ground-all-over finish. These bits can be readily identified as they are rust-proof, non-magnetic and very hard.

BLACKALLOY

This is a non-ferrous alloy of cobalt, chromium, tungsten, tantalum-niobium and a variety of other minerals in minor quantities. The composition is such that the disadvantages of low shear resistance have been eliminated, and, according to the makers, Blackalloy has outruled high speed steel on most applications and has also replaced tungsten carbide where this could not fully satisfy. They also say that the most important difference between Blackalloy and carbide lies in the entirely different manufacturing process: carbides are produced by pressing powders in moulds, and sintering, whereas the raw materials for Blackalloy are melted and cast in centrifugal casting machines. The alloy can be ground easily with wheels suitable for high speed steel, and although I have not been able to carry out extensive tests I have found that threading tools of Blackalloy hold up well. The makers, Messrs Brunner Machine Tools Ltd., of 241-7, High Street, Acton, London, W.3. supply interesting leaflets giving full particulars and technical details.

TUNGSTEN CARBIDE

Basically this may be regarded as a special alloy of tungsten and carbon, the two being furnace-fused together, after which a binding element is added followed by ball-milling to a fine powder. Hydraulic compression to moulded forms and a further semi-sintering complete the process.

A whole range of tungsten-carbide alloys falls under the general heading "carbide", and all have a similar appearance, but as experiments with different binding agents and mixture proportions are continuous, the suitability of "a carbide tip" for lathe screwcutting should not be judged from the results of trying carbide tools of unknown origin.

Sometimes trade names such as "Wimet", "Ardoloy", "Veraloy" and "Cutanit" are given to carbide alloys made to formulas especially composed to produce high performance tools for various purposes. In this way, Veraloy Products specialise in the production of formed and general purpose threading tools.

Single-point threading tools tipped with tungsten-carbide play an important part in the toolroom, although not all lathes are ideally suited to the use of carbides, which calls for rather greater rigidity of mounting and extra smoothness in the working of the lathe slides if the best results are to be obtained.

For the finest cutting edges carbide tools should be sharpened on a diamond impregnated wheel. Ordinary grinding wheels suitable for high speed steel will make no impression on tungsten carbide which is of very nearly diamond hardness, although special wheels known as "green grit" are fairly satisfactory when grinding is carried out on the periphery of the wheel. The side grinding called for when a thread tool grinding jig (q.v.) is used is very slow, and the wheel requires frequent dressing with a diamond to freshen up the surface.

CARBON STEEL

In the early days this was the only material used for lathe tools, drills, and so on. It may be assumed to consist of a good quality steel with an average of about 1 percent of carbon. Although hardened carbon steel will not withstand anywhere near the high cutting speeds associated with modern alloys, it is worth noting that solid internal threading tools such as those shown in Figs. 19 and 20 will sometimes give satisfaction when made from carbon steel or "silver" steel. Threading speeds in an ordinary lathe are seldom high, and carbon steel works quite well for brass and some of the softer grades of phosphor bronze. The disadvan-

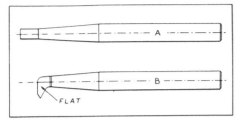

Fig. 19. *Small internal threading tools may be made by machining as at A, then cranking as at B followed by hardening and grinding to shape.*

tage of carbon steel lies in the fact that it is so easily spoiled by inadvertent over-heating during grinding.

CUTTING ANGLES

The chief terms used in describing the geometry of threading and other lathe tools are shown in Fig. 21. It seems unfortunate that the word "side" appears both in designating side rake on top of the tool, and side relief for the sides of the tool. One would have thought that it might have been preferable to refer to the (top) side rake merely as "top rake", and in those cases where the cutting edges of a tool slope backwards and downwards away from the tip (as do parting tools for steel) to refer to this as BACK rake.

As is well known, when cutting steels with high speed steel, as the side rake of a tool (Fig. 21 A) is increased, so does the traverse feeding pressure decrease: a circumstance favourable to a reduction in stresses to both machine and component. For these reasons it is sometimes an advantage initially to form a thread with a tool having a degree of side rake, and then to change the tool to the shape as at Fig. 21 B for finishing cuts.

Of course, whether or not such a method is regarded in a favourable light must be left to the discretion of the operator after a consideration of the requirements of any particular job in hand.

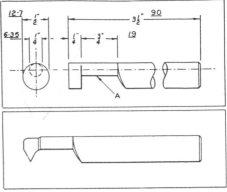

Fig 20. *An example of a 'solid' internal threading tool initially shaped by eccentrically chucking the shank and machining to reduce the diameter at A. The lower illustration gives an idea of the finished tool.*

(Underlined figures are approx. mm.)

From some points of view, the inclusion of only five degrees side rake may seem little more than a gesture in the desired direction, yet such a tool will give very good results, and for finishing cuts, the trailing cutting edge with its negative rake has been found to cut quite well.

Tools with any marked side rake for brass or bronze threading can lead to

Fig. 21. *Illustrating the terms associated with threading tools.*

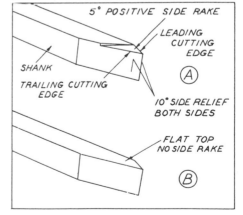

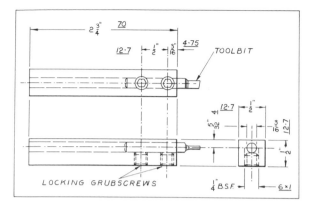

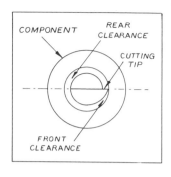

Fig 22. Details of a holder or shank for use of 3/16 in. dia. HSS toolbits.
(Underlined figures are approx. mm.)

Fig. 23. Illustrating the clearances necessary for internal threading.

complications due to a self-feeding action induced by the rake, and it is therefore generally advisable to thread these materials with a tool as at *B*, (Fig. 21)

Carbide tools are customarily used without side rake for all purposes. For Blackalloy tools the manufacturer's instructions should be followed.

An economy in tool steel can be achieved by using a bit holder similar to that shown in Fig. 22 which will take HSS bits of the ready hardened and ground-all-over type, $\frac{3}{16}$ inch (4.75 mm) in diameter and $2\frac{1}{2}$ inches (63.5 mm) or more in length.

As a matter of fact, round section toolbits of modest diameters are particularly suited for tools for square and Acme thread forms when fairly coarse pitches are required on the smaller diameters. Ordinarily, if attempts are made to grind a larger square section toolbit for square and Acme threads it is found that the necessary side relief clearances required, together with the thinness of the cutting end, will reduce the underside of the cutting end to a knife edge which vanishes at about half tool height, so that in reality a large tool

section is wasted, as is much of the time taken to grind it.

Offsetting the toolbit in the way indicated in Fig. 22 not only allows of working that much closer to a shoulder, it also gives a sufficient thickness of metal for the grubscrews to be effective and at the same time keeps the shank section to a satisfactory minimum for use in the toolposts of the smaller lathes. Toolbits for these holders are ground whilst locked in the holder, the treatment then being exactly as described for the grinding of 'solid' external tools.

INTERNAL THREADING TOOLS

'Solid' internal threading tools present certain problems of economy which do not occur with external tools. Whereas with a solid external tool the shank may be progressively encroached upon as fresh grinding becomes necessary, this of course is quite impossible for internal tools, for the reason that the cutting portion has to be at right-angles to the shank. Consequently, when an internal tool becomes past regrinding, all the bother of a fresh forging operation becomes necessary or the tool has to be

120

discarded. Nevertheless the solid type of internal threading tool is used extensively, especially for the threading of blind bores, and for this purpose two designs are offered. The first is shown in Fig. 19.

Black, round section annealed high speed steel bar is used, and a suitable length is machined in the lathe to the approximate shape shown at *A*. The tapered end is then heated and cranked as at *B,* after which the operational end is hardened. Next a flat is ground to about half diameter, and the required threading angles are formed in the manner to be described.

HARDENING HSS

The hardening of high speed steel is really best carried out by a firm specialising in this kind of work. Lists of addresses will be found in 'Machinerys' Buyers' Guide' under 'Heat treatment'. A specialist firm will give the proper treatment for any particular grade. If sending away for hardening it is as well to make up a worthwhile batch. Some firms may have a minimum charge.

HSS. WORKSHOP HARDENING

High speed steels will not give the best cutting performance unless, before quenching, heating is taken to near the melting point: i.e. a bright white heat with a temperature of about 1250 deg. Centigrade, after which tools may be quenched in paraffin – best done out of doors for safety.

Although the necessary temperature can be reached by use of any oxy-acetylene flame, it was found that the gases react in some way with the steel and render it unfit for use.

In the absence of any better heating method such as a blacksmith's forge, HSS tools can be satisfactorily hardened by using the flame of an electric carbon arc to heat the small portion to be treated, and quenching in paraffin. The writer used $\frac{3}{8}$ in. dia. copper-plated carbon rods at about 100 volts, 40 amp. AC. The HSS should be heated in the arc until 'sweat drops' appear before quenching. Coloured safety glasses must be used, otherwise the brilliance will damage the eyes. Of course this rather crude approach is suitable only for small tools such as those used for internal threading or boring, whereon a modest amount of metal may be left for subsequent grinding to shape. It was found that these small tools did not require tempering or 'letting down' after quenching.

Bending, or end-cranking for internal tools, must be done at a good bright heat.

'Black' annealed (i.e. softened for machining) round section HSS can be obtained from Sanderson Kayser Ltd., Attercliffe Steelworks, PO Box 6, Newhall Road, Sheffield S9 2SD. Weight for weight, the black annealed bar is much cheaper than the ready hardened and ground-all-over toolbits.

Once hardened, HSS is very difficult to soften by any means likely to be available in the average workshop.

HARDENING CARBON STEEL

Carbon steel or 'silver steel' (the latter known as 'drill rod' in America) may be hardened by heating until it just passes becoming non-magnetic, then quenching in water. Magnetic testing can of course be carried out by means of a small permanent magnet such as the 'Eclipse'. (The magnetic test is of no use for HSS). The writer found that small tools such as the kind under discussion do not require letting down or tempering after hardening. Indeed, the subsequent grinding to final

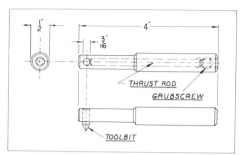

Fig. 24. Details of a bit-holder for internal threading or boring. The shank can be of square section.

shape can produce too much heat if care is not taken to repeatedly cool the tool.

The fact that internal threading tools require rather more clearance than do external tools should be noted. Generally speaking, the comparative smallness of the cutting end will take care of this aspect simply because the amount of metal below the cutting edges will itself be small. Fig. 23 illustrates the requirements which should be observed. Sometimes an initial check can be made by merely presenting the cutting end of the tool to the inside of a washer having a bore similar in size to that to be threaded: it is not then difficult to form, by observation, an opinion as to the suitability of the tool.

The internal threading tool shown in Fig. 20 has the merit that initial heating

Fig. 25. A toolbit brazed in position.

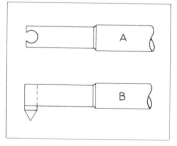

and cranking is not called for. The shank is eccentrically chucked in a lathe, and portion *A* is machined with a parting-type tool to form a space for more robust tools which can be used to complete the turning. The circular operational end is then hardened and ground down to approximately one half diameter, and the required thread angles attended to.

INSERTED BIT TOOLS FOR INTERNAL THREADING

A pleasing design for internal threading tools of the inserted bit type is shown in Fig. 24, where you will see the bit is locked by means of a thrust-rod and grubscrew.

The thrust rod may be of silver steel, with each end fairly well chamfered. Contrary to what may be thought, very fierce tightening of the grubscrew is unnecessary.

The shank may be of circular or square section. A circular section shank has the merit that it can be slightly rotated for minor adjustments of the bit position within a bore.

The initial forming and regrinding of the bits is best carried out by removing from the holder and making use of the auxiliary bit holder for the grinding jig, as will be described.

BRAZING

When a bore to be threaded is not of too small a diameter it is sometimes expedient to fix a toolbit by brazing. With this method, the absence of shank metal at the leading end makes the tool suitable for blind bore threading whilst showing considerable economy in tool steel. A tested approach to the brazing method is illustrated in Fig. 25.

At *A,* a steel shank is cross drilled to give a free fit to the bit. After drilling, about one third of the hole is faced away. At *B,* after flashing brazing metal into the hole and around the toolbit, about one third of the bit diameter is ground away at the leading end. The thread angles are then ground in the same way as will be described for solid internal threading tools.

A SIMPLE JIG FOR GRINDING AND SHARPENING

In many workshops throughout the country, lathe threading tools are shaped by freehand methods on an offhand, or bench type, grinder, and that so much good work is produced can be a tribute to the skill of the operators. However, the freehand method does call for what may be termed a series of carefully calculated guesses, especially when endeavouring to grind the thread flank angles. Repeated tests with a Vee-notched thread gauge are called for, without any guarantee that the errors thus indicated can be corrected by a fresh approach to the grinding wheel. Additional facets are likely to be formed with each new grinding after inspection. The tool heats up and burns the fingers, and, in the long run, attempts at the perfection originally envisaged are abandoned when a not-too-close inspection

seems to indicate that the tool is ready for use. By these methods the tool flanks are most unlikely to be geometrically straight, and the tool's Vee angle is even more unlikely to be symmetrically disposed with respect to the shank centre line (Fig 26A) with geometric precision, with the result that the tool has to be set in the lathe by referring to the Vee notched gauge in an attempt at getting the tool's cutting edges symmetrically positioned with respect to the component. Slight grinding imperfections again make this final lathe setting a somewhat doubtful business: one can seldom say for certain whether the shank requires swinging slightly to the left or to the right, and in the event the tool is ground and set by a series of compromises.

Admitted, threading tools can be accurately shaped on a surface grinder, but setting up on a magnetic chuck and sine table takes considerable time, apart from the fact that a surface grinder will not always be free when needed. Accordingly the hand jig about to be described is recommended. Although it is not to be denied that a little extra time is taken to set up the jig, the important aspect is that every step taken is a *positive* one towards the production of accurate threading tools.

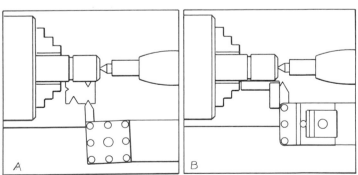

Fig. 26. A. A threading tool ground by freehand methods is unlikely to have its cutting edges symmetrically disposed in relation to the shank.

B.A jig-ground tool will be symmetrical in all respects and can be set with the larger references afforded by a small square.

Fig. 27. A grinding 'unit' with large area workrests.

The jig is used in conjunction with a grinding "unit" similar to that shown in the photograph, Fig. 27 where each wheel is provided with adjustable-angle work-rests of somewhat greater surface area than is usually offered on the smaller bench or pedestal offhand grinders. However, no difficulty should be experienced in carrying out such a simple modification, which, in any case, will be found to offer distinct advantages in the grinding and sharpening of ordinary lathe turning tools. The two wheels on the unit

Fig. 28. The thread tool sharpening jig. General arrangement.

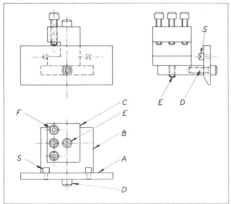

shown are of 8 inch (200 mm) diameter, and $\frac{5}{8}$ inch (16mm) face width. The work rests are each of $3\frac{1}{2}$ inch (90 mm) length, and $2\frac{1}{2}$ inch (65 mm) width, and project beyond the front of the wheels by $1\frac{1}{4}$ inch (32 mm). The drive is by means of a pulley outside the left-hand wheel, with a Vee belt to a motor mounted underneath.

Some commercially built grinders may not allow sufficient space for fitting the inner work-rests, but this will be of no consequence for thread tool grinding.

The diagram Fig. 28 shows the general arrangement, and details are given in the drawing, Fig. 29. The jig is constructed from three pieces of bright mild steel. Referring to Fig. 28, the mounting plate B is attached to the backplate A by one screw D, and the movement of plate B is limited to 10 degrees in either direction by means of stop screws S. Upon plate B there is mounted the toolholder C which is secured with screw E. The toolholder can be swivelled through a full circle. Screws F are used to lock the tool to be ground. A special auxiliary holder is used for grinding very small 'bits'.

The 10 degree angular settings of plate B relative to the backplate A determine the side relief angle of both leading and trailing sides of threading tools for external or internal use and will meet every ordinary requirement. Formally, the side relief angles favoured by the writer were 10 deg. leading and 5 deg. trailing, but of course these had to be reversed for left-hand threading tools. Further consideration led to the conclusion that as the trailing side normally has comparatively little work to do, an increase of the relief to 10 deg. at that side would have no appreciable effect on tool life, and except in those instances where top rake is desired, the same tool(s) would serve for both right and left-hand threads. Moreover, the additional relief at the

124

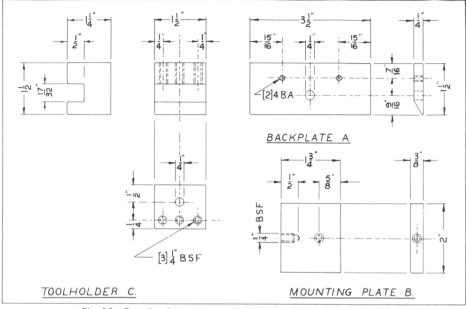

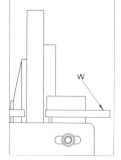

Fig. 29. Details of components for the thread tool sharpening jig.

TOOLHOLDER C.

MOUNTING PLATE B.

BACKPLATE A

trailing side favours the shaping of tools for internal threads where the clearance requirements become somewhat exaggerated. With these points in mind the construction and use of the jig is simplified.

EXTERNAL THREADING TOOL

First, the grinder work-rest is set exactly at right-angles to the side of the grinding wheel — Fig. 30. Next, a piece of straight bright steel, *L*, Fig. 31, of about $\frac{1}{16}$ inch (1.6 mm) in thickness is temporarily clamped to the work-rest *W* in such a way that the inner edge of the strip is at the exact right-angles to the side face of the wheel, regardless of whether or not the wheel has become slightly tapered in section through wear at the side. With the grinder prepared in the manner just described we may turn our attention to the setting of the tool in the jig. The necessary conditions are shown in Fig. 32.

At *A*, for a 60 deg. threading tool, the toolholder is set and locked to the desired angle by referring the setting gauge to the inner side of the backplate and the side of the tool to be ground. After setting as at

Fig. 30. The grinder workrest W is first set at right angles to the side face of the grinding wheel.

Fig. 31. Location 'strip' L is then clipped to workrest W at rightangles to wheel.

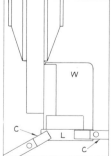

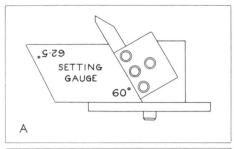

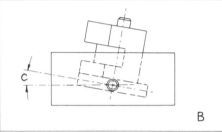

Fig. 32. At A the tool is swivelled to and locked at the necessary angle, and at B the toolholder is tilted to angle C for the necessary side relief.

Fig. 32A, the mounting plate is tilted through angle *C* Fig 32 B (limited to 10 deg. by stops not shown) for the side relief of the tool.

Fig. 33. Illustrating the principles of tangential grinding.

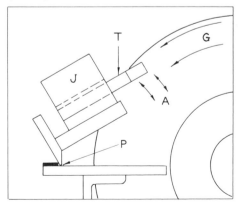

The thread angle-setting gauges are easily made from pieces of sheet mild steel of about $\frac{1}{16}$ inch (1.6 mm) thickness.

The drawing Fig. 33 illustrates the principle of actual grinding or sharpening, although it is assumed here that if a new tool is being made it will have been roughed to approximate shape by freehand grinding on the periphery of a coarse wheel. The jig is first aligned against the strip already clamped to the work-rest, and, whilst maintaining this alignment by careful hand control, the jig is advanced towards the wheel side. As soon as grinding commences, the tool is swept across the side face of the wheel through a small arc as indicated by the arrows *A* whilst the whole is pivoted about *P*. The necessary slight inward pressure required to maintain the cut is easily achieved, and when care is taken to pivot evenly against the locating strip *P*, a very accurate tool face will result. A photograph of this set-up may be seen at Fig 34. although it was not possible to include the hands, which would have obscured some of the more important details.

The direction of grinding, being of a tangential nature, not only assists in producing a perfectly flat face, it also ensures that the cutting edge is of a straightness approaching geometrical truth in addition to holding the correct angle to close limits.

For grinding the trailing, or right hand cutting side, the tool is first turned upside-down and carefully re-located in the jig, then clearance angle *C* (Fig 32B) is changed from positive to negative, whereupon the jig and tool assembly is again presented to the wheel as in Fig. 33.

By setting the tool angle only once, as in Fig. 32, then turning the tool upside-down for grinding the trailing side, symmetry of the Vee point relative to the shank is assured, assuming, of course, that the shank is of the machined or ground-all-over type.

The small amount of skill required to use the jig is soon acquired and no matter how many times the whole is removed from the work-rest to inspect the progress of the grinding, relocation in the exact position is always assured. Moreover, if the sweeping motion is continued as at A, (Fig. 33) without the application of additional in-feeding pressure, grinding will become progressively lighter and lighter, and at the same time, any irregularities in the actual wheel face will have an ever diminishing effect. That a mirror finish is achieved may be demonstrated by holding a ground face close to the eye: an undistorted image of the surroundings will be seen.

When it is necessary to grind the top face of a tool, this may be done on the front periphery of the wheel with the tool shank held to the work-rest. It is worth noting that as the shank is raised above the work-rest by the interposition of parallel packing, P, (Fig 35) so the side rake that can be ground will increase in angular value, and with this setting an angle can of course be exactly repeated.

Grinding in the manner just described does, of course, leave the tool with a sharp point: but this may be broken by means of a slip stone with the tool-shank held in a bench vice Fig 36A shows the operation. Naturally, care must be taken to hold the stone with the necessary tilt so as to maintain the front and side relief angles, but unless the tool is very small, the apex itself will position the stone, and the forefinger will readily detect any tendency to deviate or incorrectly tilt.

In the opinion of the writer (except when on quantity production) it is preferable to hold tip or apex radii to a minimum consistent with their not breaking down under cutting stresses. For quantity

Fig. 34. Grinding the leading cutting edge of a 60 deg. threading tool.

production of any one size, a good

Fig. 35. Peripherial grinding: showing how the side rake on a tool at 1 can be increased as at 2 by the interposition of packing P. Such settings can of course be exactly repeated.

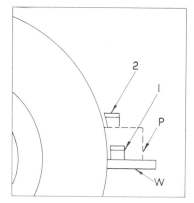

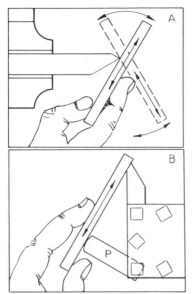

Fig. 36. At A a hone is being used to break the sharp apex of a freshly ground tool. Illustration B shows how a tool may be sharpened without removal from the lathe. Piece P may also be positioned to project outwards from the toolpost slot to the right of the tool. Please also see text.

practical approach is to commence with a minimum tip radius, then to continue depthing until the body of the thread is properly proportioned (as checked by 'wiring' – see Section 9). This of course will lead to over-depthing on the one component (or test piece). The amount of over-depthing is then read from the graduated in-feed dial and honed away from the tool apex. Thus for subsequent screws, the in-feed dial reading (read from a tool-scratching zero start) will reasonably agree with the required thread depth, and no time will be wasted by unnecessary removal of metal. This approach, of course, eliminates call for expensive tool-form checking aids such as shadowgraph projectors.

It was convenient to include with Fig. 36 the illustration *B* which shows how the leading cutting edge of a threading tool may be accurately honed when the tool is in position in the lathe. The piece *P* represents an odd length of hard steel or lathe toolbit, the free end of which has been radiused in the manner shown, and which has been ground to the same side relief angle as that of the threading tool. It is not difficult to adjust the projection and angle of the piece *P* so that it forms a guide for the hone, holding the latter to the thread tool flank face, after which, of course, both tool and guide are honed together, with a slight pressure bias in favour of the threading tool. Admitted this hones only the leading cutting edge, but usually the trailing edge has much less work to do.

Flat-top tools are of course easily honed, but without some guidance it is difficult to hone the (top) side rake because an inadvertent tilt of the hone can easily change a positive rake into a negative one. To avoid this risk the simple set-up shown in Fig. 37 may be used.

The tool *T* to be honed is held sideways in a toolmakers' vice with packing *P* and a piece of round stock *R* interposed. By adjusting the length of *P* and the diameter of *R* it is fairly easy to judge, by observation, that the hone is in uniform contact

Fig. 37. A simple set-up for accurately honing side rake on a tool.

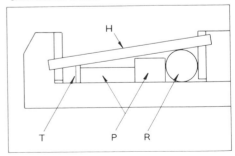

with the top face of a side-raked tool, whereupon the tool and the round stock are honed together without risk of tilting the hone.

INTERNAL THREADING TOOLS

Internal, or nut, threading tools require that the jig is used in a position somewhat further away from the periphery of the grinding wheel. The extension plate, Fig. 38, forms a convenient means of achieving this object, and, with two locating strips L1 and L2, the plate serves for both internal and external thread tool grinding. The free end of plate P is clamped to the grinder work-rest, and for internal threading tools, the whole is adjusted so that the locating strip to be used is at right-angles to the operational side of the grinding wheel in both a vertical and horizontal plane. (This plate (Fig. 38) was subsequently provided with an angle bracket on the underside, so that when necessary it can be changed for the plain workrest, thus eliminating use of toolmakers' clamps).

Fig. 38. Details of an extension plate used to aid the grinding of internal threading tools.

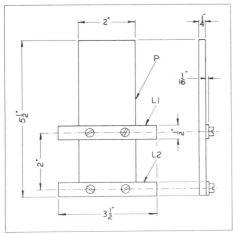

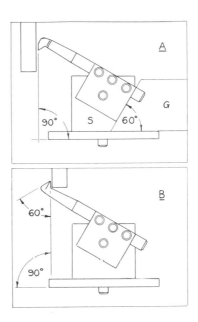

Fig. 39. The jig in use for the grinding of internal threading tools.

Fig. 39A shows how a 60 deg. internal threading tool is set in the jig, it being assumed that the toolholder slot is at exact right-angles to the end face. The mounting plate S is then tilted 10 deg. upwards at the left-hand side to set the side relief, after which the leading cutting side of the tool may be ground by gently sweeping over the side face of the grinding wheel.

For grinding the inner, or trailing cutting side, the tool is turned through 180 deg. so that the cutting end is inverted as shown in Fig. 39B. The angle of tilt of plate S is not altered, but the jig locating piece, L2 (Fig. 38) should be re-adjusted so that it is at exact right-angles to the LEFT side of the grinding wheel, or as closely as possible at right-angles to that small portion of that side of the wheel that can be reached by the tool. The jig is

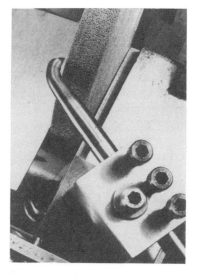

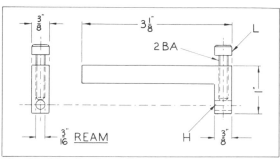

Fig. 40. Showing the trailing or inner cutting side of an internal threading tool being ground. (Left).

Fig. 41. Details of an auxiliary holder used for grinding small round toolbits. The reamed bore should be carefully aligned with the shank. (Some corners get ground away with first use of the holder).

then carefully aligned against the locating piece (L2) whilst the tool is offered to the side of the wheel as shown in the photograph, Fig. 40, whereupon a gentle movement to the right, combined with the limited sweeping motion permitted will grind the trailing cutting face. Admitted for this operation the corner of the wheel must be sharp and square, but this requirement holds regardless of whether or not a jig is used.

With tools for internally threading very small bores it is sometimes necessary:

(1) to back off the body of the tool just below the apex of the Vee, and

(2) to grind a secondary side relief of 25 to 30 degrees at the leading cutting side, but leaving the original 10 degrees relief of about 30 thou. in. (about $\frac{3}{4}$ mm) height.

Operations 1 and 2 are generally more easily carried out by a freehand approach, and for operation No. 2, with the precision outline as a guide, and by watching and maintaining a gap between the leading edge and the grinding wheel, it is not at all

difficult to grind the secondary relief without encroaching on the first precision 10 deg. angle.

SMALL BIT GRINDING

The grinding of very small toolbits of, for example, $\frac{3}{16}$ in. (4.75 mm) diameter and $\frac{7}{16}$ in. (11.0 mm) length, such as would be used in the boring bit holder shown in Fig. 24, calls for the auxiliary holder shown in Fig. 41. This is made from a $3\frac{1}{8}$ in. (80 mm) length of 1 in. (25 mm) by $\frac{3}{8}$ in. (9.5 mm) bright steel. The toolbit to be ground is held in the bore H by means of the hexagon-socket cap screw L.

After locking the toolbit in position, the first operation is to form a flat to a depth of approximately one-half bit diameter: this is done on the periphery of the wheel as shown in the photograph, Fig. 42. Note the orientation of the holder in relation to the bit to be ground. For grinding the leading and trailing cutting sides, the auxiliary holder with the bit undisturbed is transferred to the main jig, after which the sequences are exactly the same as those described for grinding ordinary solid-with-shank threading tools for external use. The photograph Fig. 43 shows the leading cutting face being ground.

ACME INTERNAL

Acme or trapezoidal form internal threading tools of the solid-with-shank type may be ground or sharpened in exactly the same way as 60 deg. solid-with-shank tools, but Acme or trapezoidal form tools of the inserted bit type require a different grinding approach. Although the auxiliary holder may be used to hold a short bit for grinding a flat on the top, as in Fig. 42, the included thread angle of only 29 deg. Acme (or 30 deg. trapezoidal) is too acute to permit use of the auxiliary bit holder for grinding the thread angles, so grinding has to be carried out with the bit locked in its own shank. Fig. 44 shows the requirements. At A, side 1 is ground in the same way as for solid-with-shank 60 deg. tools, Fig. 39B, except of course that the angle of approach is modified to Acme (or trap.) standards. For side 2, the bit is temporarily rotated through 180 deg. and the shank inverted as at Fig. 44 B, then grinding is carried out as in Fig. 45 where side D of the swivelling platform is tilted 10 deg. *downwards*.

TOOLS FOR SQUARE FORM THREADS

The jig cannot be used for grinding tools with parallel cutting sides for threads of square form. However, these tools can be shaped accurately by temporarily using a high-speed drill press as a miniature surface grinder. Fig. 46 illustrates the principle. A small grinding wheel G of the integral-with-shank type (sometimes known as 'mounted grinding points') and of about 1 in. diameter is chucked and rotated at the highest available speed. The tool, of the 'inserted bit' type ($\frac{3}{16}$ in. dia. HSS toolbit, e.g.) is temporarily rotated

Fig. 42. The auxiliary bit-holder is used for grinding a flat on a short 3/16 in. dia. toolbit.

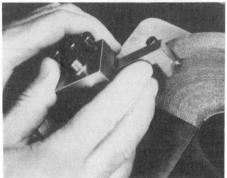

Fig. 43. Grinding a small threading toolbit for subsequent use in a shank such as that illustrated in Fig. 24.

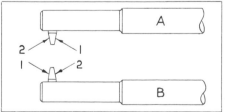

Fig.44. Above, illustrating the sequence associated with grinding an Acme form toolbit for internal threading. Please see text.

131

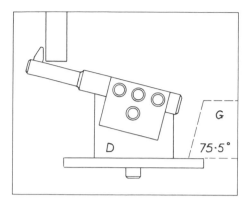

Fig. 45. Grinding an inserted toolbit for Acme form internal threading.

and locked in its shank *S* to bring the top of one cutting side face level, as at *A* whereupon the grinding wheel is gently fed down until sparks appear, while the shank *S* is held to the drill worktable *W* and gently swept about to imitate surface-grinder table movements.

On completion of side *A*, the shank is inverted and the bit again rotated to bring side *B* level for surface grinding, after which, of course, the bit is restored to its operational position with cutting edge *T* level.

The cutting end can of course be pre-roughed on an offhand bench grinder, and for easier control during surface grinding, the shank *S* is best held in a machine vice.

SQUARE FORM – INTERNAL

Tools of the inserted-bit type for square thread form cutting can of course also be ground by use of a drill press, but at the time of writing, and because of obstruction by a shank, there seems to be no ready means for the production of very accurate parallel cutting sides on internal square form threading tools of the integral-with-shank type, at least without the aid of a conventional surface grinder.

THE RETRACTABLE AND SWING LATHE TOOLHOLDER

As there are already ample descriptions of the various devices and ways in which cutting tools may be held in a lathe it is felt that no useful purpose can be served by repeating the details here. However, in the course of making the many screwcutting experiments necessary for the completion of these notes, a need was felt for something much more versatile than any

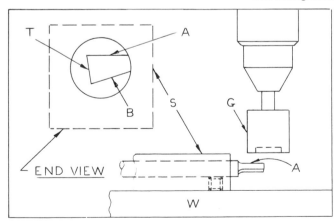

Fig. 46. Illustrating the use of a high speed drill press and small grinding wheel or 'grinding point' for the production of parallel sides on tools for threads of square form.

holder that could be found on the market at that time.

When screwcutting, considerable time is lost by the necessary repeated retraction and resetting of the in-feed, and mistakes are easily made. The situation is even worse when internally threading, because before progress can be gauged, the tool has to be first cleared from the thread, then retracted sufficiently to permit entry of a screw gauge, either by reversal of the cross slide, or by a lengthy right traverse of the carriage. Also with many toolholders the setting of a tool at the correct height can be a troublesome task, taking time that would be better devoted to cutting the thread. Moreover, with conventional toolholders, any minor height adjustment it may be felt desirable to make *after* a tool has commenced threading generally has to be postponed because of the virtual impossibility of exactly repositioning a tool relative to the helix already formed.

These numerous disadvantages therefore led to the development and

Fig. 47. *The author's general purpose, retractable, height-adjustable and swing clear lathe toolholder.*

patenting of the toolholder about to be described.

The toolholder is illustrated in the photograph, Fig. 47, where it is shown in the tool-retracted position. Turning or threading tools are mounted in the left-

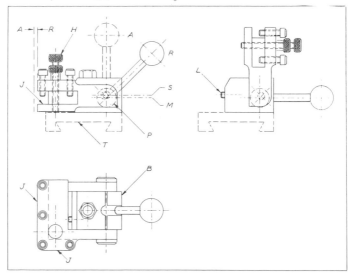

Fig. 48. General details of the toolholder shown in the photograph, Fig. 47.

hand slot, and boring or internal threading tools are held in the front slot, generally aligned with the lathe axis. When necessary the holder may be swivelled about its fixing bolt.

The construction of the toolholder is set out diagrammatically in Fig. 48 where it will be seen that the tool slots *J* are carried on arms at each side of a central body-block *B* which is fixed to the top-slide *T* by means of a bolt or stud. The arms are pivoted on bearers eccentrically disposed on centre-line *S* at each end of a main pivot pin *P*, the latter being rotatable about its centre line *M*. A ball hand-lever is screwed radially into the centre of the main pivot pin. Partial rotation of the main pivot pin by means of the hand-lever therefore operates the eccentric means, and advances or retracts the tool carrier, the two positions being shown as *A* and *R*. The main pivot-pin bore is split for adjustment by closure to a friction tightness by means of screws not shown.

The forward or advanced position of the tool carrier is determined by the hand-lever abutting the termination of its slot or clearance recess. The degree of retraction can be determined by adjustment of screw *L* which, with the tool carrier in the retracted position, abuts the back face of the front tool slot. In this example a full retraction of about $\frac{5}{32}$ in. (4.0 mm) is provided, but a movement of about 95 thou. in. (2.4 mm) is sufficient for threads

of up to 8 tpi or 3.5 mm pitch. The retraction stop *L* is however extended to limit movement when threading very small bores, otherwise when a tool is advanced to clear an internal thread, the shank will contact the rear wall of a bore.

HEIGHT ADJUSTMENT

The operational height of a tool is adjusted by means of the knurled head screw *H* which passes through a threaded bore behind the junction of the front and side tool slots and bears upon the surface of the lathe slide. The screw of course offers immediate micro-adjustment for tool height and eliminates call for the adjustment of packing. Further, when screwcutting, one frequently entertains a notion that a tool might cut a little more sweetly if raised or lowered by a minute amount, and the height adjusting screw allows of such adjustments being made between cutting passes, or, if a cut is light enough, then the height can be adjusted whilst a tool is actually cutting.

The operational height of an internal threading or boring tool is fairly easily judged by mere observation. A quick method for setting external tools to centre height is shown in Fig. 49. This approach does of course assume that a tool shank is horizontally disposed: with the type of tool height adjustment provided on the Retractable and Swing Toolholder some allowance should be made for the fact

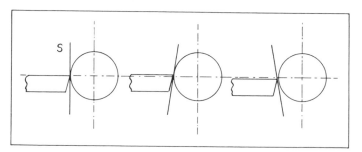

Fig. 49. Setting a tool to lathe centre height.
When a tool is correctly set, a thin metal strip S interposed between the tool and a workpiece and held by slight cross feed pressure, will stand vertical.

Fig. 50. The general purpose retractable and swing toolholder swung clear for testing the progress of an internal thread.

that a shank may slope slightly upwards or downwards.

The height adjusting screw also acts as a swing return stop to exactly reposition a tool after swinging clear for gauging. The photograph Fig. 50 shows the toolholder swung clear to test the fit of a screw gauge: an operation that can be repeated at any time without disturbing the slide settings last used.

GENERAL USE OF TOOLHOLDER

For turning and threading with a die, the swing-clear feature allows of turning to stop length, swinging clear for die running, and parting from the rear. In this way, screw or bolt heads are sized for height with only one L.H. carriage stop position, and screws may be threaded up to the head without call for excessive slide retraction to bring a turning tool clear of a die.

When two or more bores are to be machined to the same size, the cross-feed dial reading for the first sized bore may be noted, and the holder swung clear for rear tool facing, centring and pre-drilling the next component, thus no time is lost by the ordinary requirement of having to find a new slide setting for each individual bore. Ordinarily, a boring tool, or internal threading tool is "always in the way".

TIME SAVED

A series of carefully timed experiments made (1) with the retractable swing toolholder and (2) with the customary rigid tool, clearly showed that the time saved by the independent retraction facility averaged 6 seconds for each threading pass. This may sound insignificant, but in the course of lathe screwcutting over 700 feet of various threads requiring a total of about 80,000 cutting passes, the total time saved equals about 133 hours, or about three and one third working weeks of 40 hours (assuming anyone can work 8 hours a day for five days without stopping for anything). Moreover, the 133 hours saved do not include additional time saved by (1) avoidance of resetting mistakes, (2) the ability to instantly swing clear for gauging and (3) the quick tool height setting facility.

Practical Aspects of Lathe Screwcutting

SCREWS

Before each cutting pass is made, a threading tool is advanced by a certain amount to progressively deepen the thread groove. There are three chief ways of adding these successive depthings:

  (1) By advancing the cross slide only, so that the tool is depthed at right-angles to the axis of the work.

  (2) By advancing the top-slide only, this being set round to one half the included angle of the thread.

  (2A). As (2), but with the top-slide set to one degree less than one half the included thread angle, as shown in the diagram, Fig. 51.

  (3) By depthing the tool with a combined advancement of the cross slide and the top-slide, the top-slide being set parallel to the axis of the thread, as shown in Fig. 52.

Let us now examine these methods in more detail.

METHOD 1

When depthing by direct right-angular advancements of the cross slide, the tool cuts equal amounts from both leading and trailing thread flanks. The objection to this

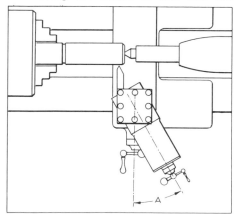

Fig. 51. The direct tangential or oblique method for in-feeding a threading tool. The top-slide is set so that angle A = one deg.less than half the included angle of the thread to be cut. For 60 deg. threads, angle A would therefore = 29 deg.

The tool is progressively depthed by advancements of the top-slide. With a forward stop for the cross-slide this may be used to retract the tool for non-cutting return passes.

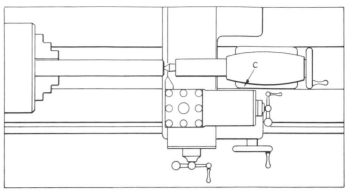

Fig. 52. If setting up to lathe screwcut with the top-slide set parallel to the lathe bed-ways and the workpiece requires tailstock support, make sure that on reaching full thread depth the top-slide will not foul the tailstock body at C.

method is that it may lead to chip wedging and flank tearing in practically all metals except brass. Further, if the tip radius or flat is not of the exact dimensions for the pitch being cut, depthing to the prescribed amount with an over-sized tip will result in an undesirable thinning of the body of the thread, as at Fig. 53 *B*, and if the tool tip is underscaled, the body of the thread will take the form shown at Fig. 53 *A*, unless the depthing is increased.

METHOD 2

This is the more generally recommended depthing arrangement. With the top-slide set round to one half the included thread angle, and assuming that the tool angle is correct and properly set, the trailing side of the tool will follow a path parallel to the trailing thread flank with the result that the leading side of the tool cuts the full depthing increment from the whole of the leading flank, and the trailing side of the tool cuts only the amount of the depthing increment from the trailing flank. Thus, with the main chip flow from the leading flank there is a reduced likelihood of chip wedging. Nevertheless, ultimate sizing at the proper depth depends upon a correct tool tip radius or tip width for reasons detailed for method (1).

METHOD 2A

This is a refinement of method (2), but here the top-slide is set round to one degree less than one half the included

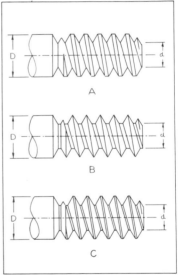

Fig. 53. Showing the importance of pitch diameter (See also Section 9).

Screws A, B and C all have identical major and minor diameters D and d, yet only C is correctly formed. At A the thread grooves require widening, and the thread at B is ruined by being depthed to the correct amount with a tool with an over-wide tip.

137

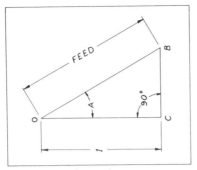

Fig. 54. When 1 (C – O) is the thread depth and the path of the tool is along the hypotenuse B – O, the depthing reading from the top-slide must be increased to compensate for the longer path. Please also see text.

thread angle: 29 deg. for 60 deg. threads, and 26.5 deg. for 55 deg. threads. The effect of the one degree less is to ensure that although with each advancement the main cutting will be done by the leading edge of the tool, the trailing edge will lightly shave the whole of the depth-length of the trailing flank: in theory, at least, because such niceties assume the perfect setting of a geometrically correct

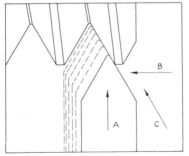

Fig. 55. Thread tool depthing with the top-slide set parallel to the lathe centre line.

When, with each depthing advancement of the cross-slide in direction A, the top-slide is advanced by a certain proportion B of the cross-slide movement, the R.H. side of the tool will follow a path C parallel to the trailing thread flank.

tool. Methods 1, 2 and 2A also have the disadvantage that an inadvertent witness mark or nick on one flank cannot be eliminated without simultaneously machining both flanks and at the same time increasing the depth.

When the top-slide is swivelled out of parallel with the cross slide, the top-slide dial readings will not give a true indication of the depthing advancements. Carried to the extreme, if the top-slide is set at right angles to the cross slide, then a top-slide advancement merely moves a tool parallel to a workpiece. However, with the top-slide swivelled as in Fig. 51 we are, in effect, advancing or depthing along the hypotenuse B - O of a right-angle triangle, as indicated in Fig. 54, and the hypotenuse holds a longer path than the direct C - O feed. Accordingly, to achieve a depthing equal to C - O, the top-slide has to be advanced by C-O multiplied by the secant of the angle to which the top-slide is slewed: C-O-B. For an angle of 29 degrees the secant = 1.1433 so if the depth of a 3.0 mm pitch ISO form thread is 1.8402 mm, the top-slide must be advanced by 1.8402 × 1.1433 = 2.1039 mm.

The accompanying Table T12 sets out the secant factors by which standard thread depths should be multiplied to

TABLE T12

THREAD	THREAD ANGLE deg.	ANGLE OF TOP-SLIDE SETTING deg.	FACTOR
Whitworth BSF	55	27.5	1.1274
		26.5	1.1174
ISO metric & UN	60	30.0	1.1547
		29.0	1.1433
Trapezoidal Metric	30	15.0	1.0353
		14.0	1.0306
ACME	29	14.5	1.0329
		13.5	1.0284

ascertain top-slide depthing readings when working to the oblique top-slide setting methods.

When threading by methods 2 or 2A without the aid of an independently retractable toolholder, it is an advantage to provide the cross-slide with a forward dead-stop so that the cross-slide can be used for tool retraction and repositioning without interfering with the settings applied by the top-slide.

METHOD 3

This is the method invariably used by the writer. With the top-slide set parallel to the axis of a thread, we have at all times a complete control over the way in which successive depthings are added.

In theory, of course, if we wish to ease the trailing cuts, for each advancement of the cross slide, the top-slide must be advanced by a certain amount, so that the trailing side of a cutting tool can be made to follow a depthing path parallel to the trailing thread flank, as indicated by the diagram, Fig. 55 at A. The net result is then similar to the oblique depthing methods 2 or 2A (Fig. 51) but with the distinct advantage that on reaching a full depthing by cross-slide advance, a series of cutting passes may be taken at that depthing, first with the top-slide retracted to clean up the trailing thread flanks as at D Fig. 4, Section 2 (if they need it) and then, as necessary, with the top-slide progressively advanced as at E Fig. 4, Section 2, not only to clean up the left-hand or leading thread flanks, but to 'thin' the thread-form to bring it to the required pitch diameter – assuming of course that the threading tool-tip is held to a minimum width as recommended in the section on tool grinding. The top-slide may be advanced or retracted in steps of 0.001 in. (0.025 mm), although because of the effect of end-play between the half-nuts and the leadscrew threads, top-slide retractions will not necessarily take effect until this play has been taken up: whereas with the top-slide advanced to cut a left-hand flank, the carriage and tool are pulled along in a positive manner, when attempts are made to shave the right-hand flanks, the leadscrew merely 'allows' a tool to traverse the thread – assuming the carriage is reasonably free to slide on the bed. However, with the foregoing process a final sizing of a thread is carried out with the least possibility of chip-wedging, and flank blemishes may be removed without over-depthing, and without being obliged to cut both leading and trailing flanks to clean up only one. As will be seen, many of these time-consuming processes are not required for quantity production. The way in which screw threads are 'wired' to gauge a correct proportioning and sizing is described in Section 9.

Referring to Fig. 55, the amount B by which the top-slide should be advanced for a given advancement A of the cross slide to ensure that the trailing flank of the threading tool follows a path C parallel to the trailing thread flank is found by multiplying the cross-slide in-feed units by the tangent of one half the included angle of the thread being cut. Thus, for a 60 deg. thread form, the tangent of 30 degrees is 0.57735, accordingly if the cross slide is advanced by one unit, the top slide should be advanced by 0.57735 of that unit: just over one half. For 55 deg. thread angles, the tangent of $27\frac{1}{2}$ deg. is 0.52056, somewhat nearer to one half than tan. for 30 degress. However, in practice every satisfaction is given by taking both tan. figures as one half, and for 55 deg. and 60 deg. threads the top-slide may be advanced by exactly one half the depthing increment added by the cross slide. For Acme and trapezoidal threads an advancement of one quarter that applied to the cross slide will serve.

When screwcutting the harder metals by method (3), it is advisable to make the appropriate top-slide advancements for each cross slide depthing increment fairly carefully until the thread is nearly completed, after which it is permissible to make final depthing increments of not more than 0.001 in. (0.025 mm) with the tool cutting both flanks, chiefly to clean up the trailing flanks. (A too-pedantic following of top-slide advancements can leave a ragged trailing thread flank). Then, if the thread is still oversize to gauge, and the depth is known to be correct within the limits laid down (as read from the cross slide feed dial previously set at zero with a 'tool scratching' start) the body of the thread may be 'thinned' by taking passes with the top slide progressively advanced thou. by thou. until a gauge fits, or a 'wire' measurement (q.v.) shows the thread is properly proportioned.

We should also note that when only one or two components require a lathe screwcut thread — components that have perhaps already had a lot of other work carried out on them — there is nothing to be gained with attempts at threading at the highest speeds: one should proceed with circumspection.

With soft or 'crisp' metals such as brass which will not easily tear, top-slide advancements seldom make any difference to the finish or sweetness of the cuts, and a tool may be directly depthed by the cross slide.

QUANTITY PRODUCTION

Meticulous attention to top-slide advancements, however, can often be dispensed with when on quantity production in free-cutting or 'leaded' steels. There is the point, though, that when a batch of the same thread is to be lathe screw-cut, one cannot commence immediately at the highest speeds, one has to work up to these, judging from an observed behaviour of the cutting passes. Once satisfactory speed, depthing increments and top-slide adjustments have been found, it is as well to make a note of them for future use. But even with the aid of these figures, for a new batch it is generally necessary to complete ten or so threads with frequent cross feed dial adjustments before the threads commence to finish to gauge size at the dial settings previously noted. It is virtually impossible to reproduce previous conditions exactly. For example, if a previous batch of threads proved to be on size at cross feed dial reading 44, and, because of tool sharpening or whatever, it was found that from a zero tool-scratching start, depthing had to be continued to say 47, then on completion of that thread I would reset the cross feed dial to read 44, then carry on working to the settings noted for the previous batch. In the event of it being thought that this approach might lead to over-depthing, then a slip-stone would be used in situ to remove a few thou. in. from the apex of the tool. Of course, by the time the ninth or tenth thread is finished to size, one will have again memorised both cross and top-slide settings.

PRACTICAL EXAMPLES.

Some examples from the author's practice will now be given. The threads are required in batches of 100-150 each. All are in free-cutting mild steel, screwcut with ordinary HSS tools, jig-ground with flat tops (no rake). The independently retractable toolholder is used, and the lathe is fitted with the single-tooth dog-clutch control for the leadscrew drive.

Non-cutting return passes are made by the pull of a spring up to a right-hand carriage dead-stop, as shown in the photograph, Fig. 55A, and, for the longer screws, depthing increments and top-slide adjustments are made whilst the carriage is being pulled back to its right-hand stop. (For screws of greater length, a cord, pulley and weight would offer a more uniform pull for carriage return.) For the 'spring return' fairly thick oil is smeared over the lathe bed: this slows the return traverse and prevents the carriage from slamming into the right stop. Soluble oil and water is applied with a brush, and the excess is absorbed by a piece of sponge positioned on a travelling chip tray. When the sponge is saturated, the excess is squeezed back into the coolant pot. This procedure ensures that the lathe carriage slides exclusively on oil as distinct from oil and water. The times given in the following examples are the times AFTER satisfactory settings have been 'found', and include loading, gauging and unloading.

necessary with a fair regularity it may indicate that the tool has commenced to lose its keenness.

With the present more common type of lathe, one would be prevented from working at these speeds because of the difficulty of having to manually arrest carriage traverse on completion of each cutting pass, and the tendency to 'snatch' when attempting to re-engage the half-nuts on to a fast-revolving leadscrew. Also, in the case of the 13 and 14 tpi threads, one would have to wait for the leadscrew indicator to register favourable half-nut engagement moments for every cutting pass. However, there is no reason why the depthing increments themselves should not be used as a guide for threading free-cutting steel at a slower and more manageable rotational sped. For harder steels, of course, the magnitude of each depthing increment should be reduced. It sometimes pays to take an additional threading pass at the setting used for the previous pass so as to check whether or not an excessive strain is building up.

(1). 1/2 in. × 13 tpi UNC, 1 in. lengths. Speed 325 rpm. 55–70 seconds.

Cross-slide	20	30	40	42	45	47	48
Top-slide	0	5	10	—	—	—	—

(2). 7/16 in. × 14 tpi UNC 1 3/8 in. lengths. Speed 420 rpm. 90–110 seconds

Cross-slide	20	30	35	40	42	43	43.5	44
Top-slide	0	5	7½	10	11	—	—	—

(3). 3/8 × 16 tpi UNC. 3/8 in. lengths. Speed 325 rpm. 35–50 seconds

Cross-slide	20	30	35	37	38
Top-slide	—	—	—	—	—

(4). 5/8 in. × tpi Acme form. 1 5/8 in. lengths. Speed 325 rpm. 120–150 seconds

Cross-slide	25	35	45	53	60	64	67	70–71–72–73
Top-slide	0	2	2	2	1	1	—	—————

These threads are sized to a hardened and ground-thread ring gauge, one for each size. Occasionally a gauge will not fit until another cutting pass has been made at the same 'last' setting. If this becomes

Please see also the notes on page 145. Ideally a thread gauge should fit with a slight 'drag' although some latitude is permissible. If for any reason a gauge is thought to fit too freely, then the

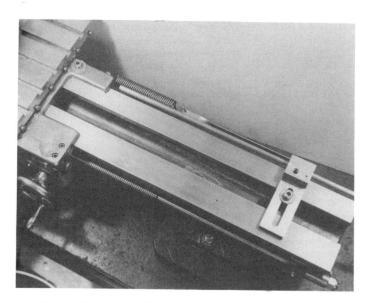

workpiece is rejected. One per-cent may be rejected for this or other reasons. Sometimes a workpiece will slip in the chuck - with disastrous results to the thread. This, of course, is entirely the fault of the operator in not adequately tightening the chuck.

NOTE: When lathe screwcutting there is always some tendency for the workpiece to be pushed away by the cutting stresses at the commencement of each cutting pass (even with fairly large lathes I am told) although of course this action is greatest during the heavier roughing cuts, but quite often the first few turns of a thread will remain oversize, even after many fine finishing passes. Often if this is not taken into account, and threading is continued in a straightforward manner until a gauge fits nicely over the first few turns, the gauge will spin freely over the remainder of the thread. This effect can be countered when a thread is finished (or is thought to be finished) by deliberately over depthing the first two or so thread turns, then smartly withdrawing the tool. An over-depthing of 0.002 in. (0.05mm.) is usually sufficient.

Batches of similar metric threads are also called for: 12.0 mm. dia. X 20.0 mm. pitch, 11.0 mm. dia. X 1.75 mm. pitch, 10.0 mm. dia. X 1.5 mm. pitch, and 16.0 mm. dia. X 4.0 mm. pitch, all except the last being of ISO form 60 deg. The 4.0 mm. pitch is trapezoidal (similar to Acme, but with a 30 deg. included thread angle.) The trapezoidal thread is somewhat deeper than the 8 tpi Acme, therefore more cutting passes are required.

A SPECIAL THREAD.

On one occasion it was necessary to cut Holtzappfel threads on three adaptors for fitting Holtzappfel face-plates and similar to the nose of a Myford wood-turning lathe.

Fortunately a Holtz. tap was provided for use as a pattern. The thread angle was found to be 45 deg. included, and the tap had very nearly sharp thread crests and roots, consequently it presented a somewhat strange and unconventional appearance. However, after much deliberation with an eye-glass and a metric rule, the pitch was found to be nearer to that of 2.7 mm. than any other (English or metric), and with this pitch, and had the root of the thread been fully sharp, the thread depth (by trig.)would-have been 3.259 mm. (0.1283 in.). As the tap had five flutes, the only way of measuring the major diameter was to bore a collar, bit by bit, until the tap pushed in, then measure the bore. By these means the major diameter was found to be 20.57 mm. (0.810 in.).

For the Holtz. thread of 2.7 mm. pitch, a leadscrew of 8 tpi was geared:

$$40 - A - 35$$
$$32 - 43$$

Before cutting the threads on the workpiece upon which a lot of work had already been carried out, including the internal threading 1 in. X 12 tpi, a thread ring gauge was made with a bore of 15.69 mm. (0.618 in.) which was screwcut progressively until the tap could be felt to enter under modest torque, after which the embryo gauge was removed from the lathe and finish tapped with the Holtz. tap.

Next, with the aid of the Holtz. thread gauge, a practice Holtz. thread was cut on an odd piece of stock of the same kind as that used for the components proper: free-cutting mild steel. From the facts available, a major dia. of 20.32 mm. (0.800 in.) seemed appropriate, with a thread depth of 2.54 mm. (0.100 in.), and having reached that depth, the thread pitch dia. was brought down in stages by top-slide advancements until the gauge screwed on nicely. This dummy run clearly showed that because the thread was so deep, and the angles so steep, it was necessary to keep top-slide advancements fairly well

ahead of proportionate depthing increments, then to continue to add small depthing increments until the depthing caught up, so to speak, and the tool again shaved the trailing thread flanks. Without this precaution there were distinct signs that a dig-in would result in spoiled work. Signs of adverse cutting action are (1) A visible lifting of the component as it tries to 'ride up' the tool, and (2) if a workshop is otherwise quiet, additional noise will be heard: sometimes an extra 'growl' from the change gears.

As a result of the test run it was felt prudent to thread the three components at 95 rpm with a series of fairly small depthing increments, consequently each took 30 minutes to complete although of only 19.0 mm. (3/4 in.) length, and despite the aid of an independently retractable toolholder and single-tooth dog-clutch to eliminate pick-up problems. Admitted, this time does not contrast at all favourably with times for the four examples of quantity production, but on the other hand, all three threads were completed to a blemish-free finish and nice fit to gauge, which for only one or a few threads is really all that matters.

NOTE: A soft steel 'nut' should never be used as a gauge for quantity production. Ordinary commercial nuts are almost invariably grossly oversize, and any 'special' nut or gauge, even if threaded with a 'good' ground-thread tap, will be worn oversize after very little testing on nearly finished tight screws. In any case, ground thread taps have to be made slightly oversize to permit entry of any screws that may be threaded to a full basic size.

THREAD CREST BURRS.

Returning to quantity production. With the smaller screw threads, sometimes the tool will leave neglible crest burrs, and sometimes a touch with a fine file is required. With the coarser threads such as the 8 tpi Acme form and the 4.0 mm. pitch trapezoidal, quite pronounced burrs are almost invariably thrown up, particularly during the fairly severe initial depthing cuts, and these burrs can be of a nature that a fine file makes little impression. Accordingly, to remove these crest burrs, a wide flat-nose (with flat top, no rake) tool is positioned in a rear toolpost, and just before the last two or three finishing cuts, this rear tool is brought forward to a previously noted dial reading and passed over the crests by one single screwcutting traverse movement (from the leadscrew, at screwcutting speed) thus removing the burrs and leaving a fine bright finish over the $1\frac{5}{8}$ in lengths in about 2.4 seconds, after the initial setting.

USE OF TRAVELLING STEADY.

As with ordinary turning, a component of any significant length or projection from a chuck should be given tailstock support when screwcutting. Sometimes for longer screws the use of a travelling steady is called for, although this should be avoided whenever possible because of the inconvenience caused by the burrs that are often raised above the work surface, especially when threading steels. When a travelling steady is used, one has no option but to file off the burrs between each cutting pass. (Although many writers apparently feel it would be demeaning to say so). Sometimes, however, after a thread has been fairly well shaped, the tendency to burr formation from subsequent lighter cuts is minimal. The possibility of incorporating a flat-ended cutting tool with a travelling steady with the idea of removing crest burrs immediately after formation has been considered but not tried by the writer.

144

SQUARE THREAD CUTTING.

That the customary text-book approach to the cutting of square thread forms is to use a parallel sided tool having a width of one-half pitch, and to depth by direct increments of the cross-slide, seems to indicate that the method will give satisfactory results despite the fact that the necessarily slender nature of the tool renders it easily prone to breakage from chip wedging. However, apart from that consideration, when a tool is of full half-pitch width there is no possibility of giving the thread flanks a final shaving without thinning or undersizing the thread body.

Accordingly, if a square form thread is to be cut to a good finish with a minimum risk of tool breakage it is worth considering the steps outlined in Fig. 56 which shows, at A, an initial and partial depthing with a Vee form threading tool followed at B by a final depthing with a parallel sided tool with a width of less than one-half pitch, and without top rake, the tool being used as at C and D for alternate flank shaving and final proportioning. With the superior strength of the Vee tool, the operation at A should be completed without trouble, although care must be taken to hold the crests to just above half-pitch in width.

It is sometimes an advantage to shave the flanks, in turn, with separate miniature knife-type tools, with each cutting side set at right-angles to the thread axis by means of a small square observed through a watchmaker's glass. With individual left and right handed flank-finishing tools it is possible to take advantage of the superior cutting qualities afforded by top rake when cutting steels.

The foregoing was the method adopted by the writer when making his square thread-form leadscrews which were cut in EN8 - a carbon steel nearly as hard as silver steel 'as bought'.

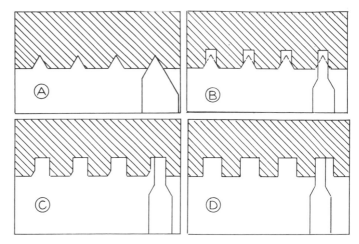

Fig. 56. It is sometimes an advantage to initially rough a square thread with a Vee threading tool better able to withstand heavier cutting than a comparatively slender square thread tool.

It will be understood, then, that the square thread form can be somewhat troublesome to produce with the certainty of a blemish-free finish. Fortunately however, a corresponding nut thread will probably be cut in a material such as cast iron which is less prone to flank tearing,, although direct depthing with a narrow tool and bringing the thread form to correct width by alternate flank shaving with top-slide adjustments is recommended.

If a square form nut thread has to be cut in steel, then it is an advantage to use a narrow tool and to alternately advance and retract the top-slide (this being parallel to the lathe bedways) at each depthing increment. Then, on reaching full depth, to shave the whole of each flank in turn by further top-slide adjustments. Of course, with this method one has to keep very careful track of the top-slide movements, the total being just short of the difference between the thread groove width and the width of the (narrow) tool. One must also of course take into account any end-play between the leadscrew thread and half-nuts.

If more than a few square thread nuts are to be processed, then of course it pays to make or buy a tap for quickly sizing threads nearly finished in a lathe.

ACME THREADS.

The Acme (and trapezoidal) form threading tool is marginally more robust than a square thread form tool of corresponding pitch, and, as we have seen, the Acme form can be cut without a change of tool. However, for the harder steels it is advisable to use a tool having a tip-width

TABLE T13

Threads/inch	Root width (max.) Inch.
16	0.023168
14	0.026478
12	0.030891
10	0.037070
8	0.046337
6	0.061782
5	0.074140
4	0.092672
3	0.123566
2	0.185350

145

fractionally less in width than the root of the pitch being cut, then, on reaching full depthing, the thread form may be brought to correct pitch diameter by top-slide advancements. (Top-slide set parallel to bedways).

As a guide, the accompanying Table T13 gives maximum root widths for a selection of Acme thread pitches.

NUT THREADS.

It appears that in the early days of Whitworth threads it was common practice to size a nut bore (minor diameter) by deducting twice the screw depth from the nominal or major diameter of the screw with the object of obtaining a near 100 percent fit between screw and nut threads: an approach from which considerable trouble must have arisen from interference between nut thread crest

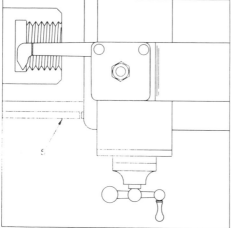

Fig. 57. When internally threading a blind bore on a lathe without an automatic disengaging device to arrest carriage traverse it is convenient to set a stop-rod S to give visible indication of the permissible limit of travel. Of course the carriage must not be allowed to actually contact a rigid stop. (See also Fig. 16 for details of a spring buffer-type 'stop indicator').

machining-burrs and the root of the mating screw or gauge. However, it is now recognised that 100 percent nut threads are unnecessary, and that the percentage reduction in nut thread strength is very much less than proportionate to a corresponding percentage reduction in the thread depth. Modern practice therefore is to recommend minimum minor diameters which offer a reasonable working clearance between nut thread crests and screw thread roots. In the smaller screws, these clearances are also more suited to direct tapping, there being less tendency for taps to clog and break.

From the foregoing it will be understood that the design dimension of a nut thread depth is now less than that for a corresponding screw thread because the basic nut thread depth is taken from the surface of the enlarged minor diameter to the design radius of the corresponding screw.

Except with special machines, direct tapping without a previous roughing in the lathe is seldom possible on work of any significant size. For example, to tap a thread of about 2 inches (50.0 mm.) diameter and 10 T.P.I. (2.5 mm. pitch) in steel requires about $3\frac{1}{2}$ horse power, which at 50 R.P.M. shows a torque of roughly 4400 inch-pounds. On the other hand, such a thread could be cut easily by a number of passes with a single-point tool in a small lathe driven by a motor of only $\frac{1}{4}$ H.P.

Despite the foregoing circumstances, it also appears that many draughtsmen assume that all internal threads will be directly tapped: a detail to be inferred from drawings which often show a nut thread terminating abruptly at the base of a blind bore. Although a lathe with a dog-clutch leadscrew control will thread to within a very small distance of a base, and it is not necessary to especially form a runout clearance, a majority of today's lathes call for the pre-machining of a runout recess, (Fig 57) and when this is not shown on the drawing, and the work is by sub-contract, it is necessary to telephone for permission to form one. This could be avoided by always drawing a dimensioned recess, and, if it is optional, adding a note to that effect. In this way all possibilities would be covered.

Internal threads carry the same nominal size designations as the corresponding screws: the apprentice should therefore remember that the minor diameter (bore size) will be unlikely to appear on the drawing. Until one has become accustomed to this it is fairly easy to make a mistake by boring the nut blank to major screw diameter, with unfortunate results. In this respect, too, it would be an advantage if draughtsmen always stated the permitted minor diameter and offered limits instead of leaving it to the turner to calculate. The draughtsman is in the best position to know what he wants, and will have all the necessary references to hand in a quiet office, whereas the turner will be working in a noisy and comparatively dirty place where he would be glad to keep reference hunting and calculating to a minimum.

Simple formulae for calculating screw and nut thread depths, nut minor diameters (bores and tapping sizes) are given in Section 1. Let us now see how these thread sizing formulae will serve in practice. We will assume that an ISO Metric screw and nut is required, the screw being of 27.0 mm. diameter, 3.0 mm. pitch, to be machined on a lathe with English feed dials.

Dia. of bolt in inches = 27.0 × 0.03937
= 1.06299 inch (say 1.063 inch, plus 0.)

Depth of bolt thread
= Pitch (mm) × 0.0241 = 0.0723 inch.

Bore of nut (minor diameter) equals:

Major screw dia. (inch) minus Pitch (mm.)
  × 0.0426

So bore = 1.063 minus 0.1278
= 0.9352 inch (say 0.936 inch, minus 0)

Depth of nut thread (minimum)
= Pitch (mm.) × 0.0213 = 0.0639 inch.

The calculated basic depth of a nut thread shows the minimum depth which will bring the nut thread roots to major screw radius when the depth figure is read from the inner surface of a recommended minimum nut bore. Slight additional depthing is therefore called for to ensure that the nut thread roots will clear the thread crests of a corresponding fully sized screw.

Example 2. What is the minimum minor diameter of a nut for a Unified thread of 8 T.P.I. and 1.0 inch diameter

Minor dia.

$$= \text{Major screw dia. minus} \frac{1.0825}{\text{T.P.I.}}$$

$$= 1.0000 - 0.1353$$

$$= 0.8647 \text{ inch.}$$

And (2A) the same nut working by millimetres:

Minor dia..

$$= \text{Major screw dia. (mm.) minus} \frac{27.5}{\text{T.P.I.}}$$

$$= 25.4 - 3.4375$$

$$= 21.9625 \text{ mm.}$$

The metric minor diameter here agrees with the inch figure above by plus 0.00088 mm. (about 34 millionths of an inch).

It will be interesting to note here that the basic thread depth for a Unified SCREW of 8 T.P.I. is 0.0766 inch (1.947 mm.), a one hundred percent nut thread would therefore require a bore of 1.0000 inch minus (2 × 0.0766) = 0.8468 inch (21.508 mm.). The formula above gives a minor diameter of 0.8647 inch which exceeds the 100 percent thread bore by 0.0179 inch (0.45466 mm.) Accordingly with a centralised mating screw of basic size the recommended minimum minor nut diameter offers an annular clearance of 0.0179/2 = 0.00895 inch, say 0.009 inch (0.2286 mm.) and nut crest burrs raised during machining would be unlikely to interfere seriously with size testing.

PERCENTAGE APPROACH.

Although minor diameters for nuts have been standardised to a recommended minimum, a closer matching of nut thread crests to screw thread roots may be made by the percentage approach if a more refined fit is felt desirable. The formula giving nut minor diameters in terms of the percentage of full screw thread engagement reads:

Minor (bore) dia.

$$= \text{Major screw dia. minus} \left(\frac{2d \times \% \text{ req.}}{100} \right)$$

where d = the standard basic thread depth of the screw, and % req. = percentage of thread engagement required.

As an example of the use of this formula, let us suppose that we require to thread a lathe backplate to fit a spindle nose having a diameter of 35.0 mm, and a pitch of 4.0 mm. (ISO Metric) and that 95% **thread engagement** was felt desirable: to **what diameter** should the backplate be initially bored?

$$\text{Bore} = 35.0 \text{ mm. minus} \left(\frac{2 \times 2.4536 \times 95}{100} \right)$$

$$= 35.0 \text{ mm.} - 4.662 \text{ mm.}$$

$$= 30.338 \text{ mm.}$$

The figure 30.338 mm. is larger than the bore required for 100 percent thread by 0.2452 mm., so there will be an annular

nut-crest to screw-thread-root clearance of 0.12 mm., (0.0047 inch).

The standard recommended minor diameter is 30.67 mm. which would give an annular clearance of 0.2886 mm., (0.01136 inch): approximately twice that given by the 95 percent calculation.

As a further example: to what size should a nut blank be bored to give 95 percent for 6 T.P.I., 1.500 inch diameter?

$$\text{Bore} = 1.500 \text{ minus} \left(\frac{2 \times 0.1022 \times 95}{100} \right)$$

$$= 1.500 - 0.1942$$

$$= 1.3058 \text{ inch.}$$

ACME NUTS.

Minor dia. = Major dia. minus pitch.

For example: What is the minor diameter of a nut to suit an Acme screw of 0.875 inch diameter, 8.T.P.I.?

Minor dia. = 0.875 − 0.125

$$= 0.650 \text{ inch.}$$

The standard recommended major diameter of an Acme nut is, for all sizes, equal to:

Major screw dia. plus 0.020 inch.

However, in the absence of specific instructions, satisfaction would no doubt be achieved by depthing the nut thread to somewhat less than the additional 0.010 inch and then widening the thread groove by degrees until the screw enters with a reasonably satisfactory feel.

For some applications a white metal nut is cast around the screw: a method which, of course, is capable of giving an excellent all-over fit.

CENTRALISING ACME THREADS.

Briefly, this thread form is such that screw and nut thread crests contact screw and nut thread roots before the mating thread flanks are in full contact, thus preventing a tendency to flank wedging, which, because of the sloping flanks in the Acme form, can sometimes give trouble.

SQUARE THREAD NUTS.

The minor diameter of a square-thread nut varies with the pitch, and the major diameter of a nut (depth to which thread is cut) may exceed the major diameter of the screw by any amount felt reasonable to allow the screw to revolve without bind and with a lubrication clearance.

NUT THREAD DEPTHING

There are five possible arrangements for progressively depthing a nut thread and these are here presented in descending order of popular usage:

(1). The tool is mounted for working in the conventional way with the cutting edges uppermost, and depthing increments are applied by successive retractions of the cross slide, Fig. 57.

(2). The tool is mounted as in (1), but with the top slide set parallel to the lathe bed. Depthing is by successive cross slide retractions, and the trailing cutting side of the tool is made to follow a path parallel to the trailing thread flank by appropriate top-slide advancements (similar to Method 3 for external threads).

(3). The tool is inverted so that it cuts at the rear of the nut, and depthing is made by advancing the top-slide which is set round to one half the included thread angle, A, Fig. 58 (or to one degree less than one half the included thread angle).

(4). The tool is inverted as in (3), but depthing is made by direct advancements of the cross slide.

(5). The tool is inverted as in (3), but held with the top-slide parallel to the lathe bed so that depthing may be made by advancements of the cross slide together with top slide advancements to ease the trailing cut as with Method (2).

Methods (1) and (4) have the objection that the tool cuts equal amounts from both leading and trailing thread flanks, although this is not such a serious consideration when cutting non-ferrous metals.

Method (3) has the advantage that only one slide setting is required in a positive direction for each depthing, and trailing cutting relief is assured.

Methods (2) or (5) offer the greatest control over depthing and final sizing. However, Method (5) cannot be adopted if the author's Independently Retractable Swing Clear Tool Holder is being used because the holder requires a downward pressure on the tool. On the other hand, depthing by cross slide retractions with independent tool retraction practically eliminates all possibility of error because the tool is cleared for non-cutting return passes without loss of the slide dial

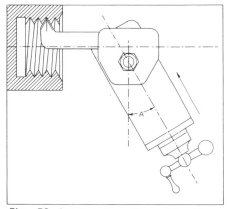

Fig. 58. Internal threading with diagonal depthing. The workpiece is rotated in the normal direction, but the tool is inverted and cuts at the rear side of a bore.

reading. Accordingly, Method (5) could be recommended when a rigid toolholding device is being used, and Method (2) has distinct advantages with the Retractable Swing holder because progress testing can be made at any time without altering the slide settings.

When directly depthing by Methods (1), (3) or (4), care must be taken to see that the tool tip radius is either correct or does not exceed the root dimension for the pitch being cut, otherwise when the nut thread has been depthed to a sufficient degree to admit the major screw diameter, the body of the thread will be undersized. On the other hand there seems to be no objection to using a tool with a minimum tip radius consistent with its not breaking down, and to continue depthing increments until the major and pitch diameters of the nut are of a size to permit comfortable entry of the screw.

With Methods (2) and (5), and using a minimum tip radius tool, a rather more careful approach to depthing is called for and it is as well to pre-machine a very short bore having a diameter fractionally larger than the major screw diameter: depthing is then continued until the tool commences to scratch the surface. Then, with this assurance that the major nut diameter is satisfactory, the pitch diameter of the nut may be attended to by flank shaving with top slide adjustments in the same way as for external threading with a parallel top slide. On completion of the thread, the witness bore may be chamfered away. This approach is particularly satisfactory for square and Acme threaded nuts, and for Vee threads in the coarser pitches.

Admitted there are objections to pre-machining a witness bore to indicate thread depth, because it calls for an additional operation. On the other hand, if a number of parts are to be threaded, then

the dial reading may be noted for the first satisfactory depthing, after which the remainder of the components can be depthed to that reading. Attempts at directly depthing a nut thread by reading the feed dial are not always satisfactory (for Methods (2) or (5)) because of uncertainties introduced by tool-spring and permitted variations in minor (bore) diameters, and if for any reason it is found necessary to change a tool when a thread has been partly depthed, then subsequent attempts at depthing from the feed dial are little better than guesswork. If a nut thread is inadvertently underdepthed, then of course no amount of subsequent flank shaving by top-slide adjustments will permit entry of the screw or gauge.

It is generally preferable to use a plug gauge to size nut blank bores if only for the reason that subsequent tests with the gauge will leave no doubt as to whether or not thread crest burrs are interfering with entry of a screw gauge.

Crest burrs can be removed with a hand scraper, although if an arrangement similar to that shown in the photograph, Fig. 59, can be adopted, then the original boring tool can be swung back into position and used whilst the threading tool is temporarily swung clear. And, of course, both holders can be swung clear to permit entry of a gauge: a useful feature, especially if the testing screw is integral with a long shaft.

A stiff toothbrush is useful for the removal of debris prior to testing.

TAP FINISHING

When a fairly large number of components are to be internally threaded, and the design of the component is such that is could not be gripped with sufficient security for direct threading with a tap, and where it is important that the thread should be symmetrically disposed about

Fig 59. A rear swing boring toolholder in use for removing nut thread crest burrs.

its axis, it is not unusual to finally size the thread with a tap after lathe screwcutting to within about 10 percent of full thread. A tap, of course, sizes the thread at one single pass and thus saves a considerable amount of time otherwise needed to achieve a satisfactory fit by successive passes with a single-point tool.

Special taps for internal thread finishing are sometimes made in the toolroom to avoid lengthy delivery times. In these instances it is common procedure to thread the tap as a copy of the corresponding lathe-cut screw, ensuring only that all basic maximum dimensions are fractionally larger than the corresponding screw dimensions. For example the major diameter and pitch diameter (see Section 9) of a tap thread may be machined to plus one or two thousandths of an inch (plus 0.25 mm. to plus 0.05 mm.) with the thread depthed from the surface of the enlarged diameter.

When the call for tap making is infrequent, and regular tap-fluting milling cutters are not available, quite good results can be obtained with a plain milling cutter which will flute a tap in the way shown in Fig. 60. After hardening, the exposed thread profiles can be dressed and sharpened on the side of a grinding wheel. The writer has also fluted taps with a suitably end-radiused fly-cutter.

OBSERVATIONS
Thread Crest Radii

The writer's own views on the matter of crest radii for the Whitworth form thread are that far too much emphasis is placed upon this feature: far more so, incidentally, by amateurs than by general engineers. Personally, if a drawing calls

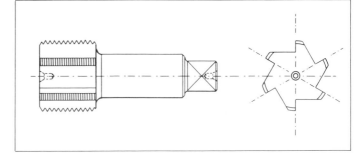

Fig. 60. A tap for finish sizing lathe screwcut internal threads.

151

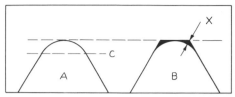

Fig. 61. Illustrating the practical implications of omitting the crest radius of a thread.

For a Vee thread of 10 tpi, dimension X is about 0.002 inch. For 2.5 mm. pitch, X = about 0.05 mm. Please also see text.

for a thread of Whitworth form, and it is to be lathe screwcut, I always put the question: 'Will flat crests be accepted?' An answer in the negative has yet to be given, but at any suggestion that this feature was of importance, a request would be made to be excused from quoting for that particular job. There are many establishments where the production of accurate thread forms is a speciality, so why not let one of those firms do the work?

First impressions are that omission of the radius would result in loss of metal down to line C, as shown in Fig. 61 at A, whereas in fact omission of the radius merely leaves the additional blackened-in metal shown at B, where dimension X, for a thread of 10 tpi for example, equals only about 0.002 inch, or approximately Pitch X 0.02 – which some turners like to try to remove with a strip of carborundum cloth or a fine square section file when a thread is not too coarse. However, a screw will be unlikely to be (indeed need not be) of full basic major design diameter, and with each 0.001 in. reduction of radius of a screw, the effect of crest radius omission becomes less pronounced. Also there appears to be no standard specifying that crest radii should follow a thread form down to its lowest permissible major diameter limits.

The objection to attempting to produce thread crest radii with a multiple-tooth chaser after sizing with a single-point threading tool lies in the virtual impossibility of so fixing the chaser in the lathe toolpost that it will be in exactly the correct position under operating conditions when gearing backlash and cutting stresses have been taken up. Also any thread to be so treated must either be free of shoulders, or must have a runout groove of sufficient width to allow the whole length of the chaser to traverse fully clear of the thread.

Some text books advocate thread finishing by use of a hand-held chaser and simple tool-rest: a method whereby the chaser teeth automatically fall into the correct position on a nearly finished thread, although personally the writer would class the use of chasers as 'messing about'. Moreover such fiddling could add greatly to the cost of threading, even assuming hand chasers are available today: and what of the corresponding nut thread, is one to fiddle with that too?

There is the odd occasion when one might allow oneself to be persuaded to try to put in crest radii against one's better judgement, and on the basis of a kind of loose assertion that although crest radii are desirable, they are not really important. In business such an approach could prove costly. One might fiddle about and produce two or three threads that satisfy a customer - who may then order a batch of 100 off; whereupon one is committed to seemingly endless fiddling with no guarantee that all will be accepted, or indeed that in the finishing operations one will not be obliged to reject some of one's own work.

Sometimes a nearly completed lathe screwcut thread is finished with a button die. The disadvantage here is that a die can remove more metal when backing off, sometimes even after opening out the die to make it slack. Again, if a thread of any significant length is finished with a button die there is a risk that the die will zig-zag its way along, cutting more deeply first at one side, then at the other, and the anticipated perfection will not be achieved.

In general therefore, it is wise to quote only for what can be done efficiently as distinct from hopefully.

Admitted with sufficient runout clearance, soft metals such as brass can be lathe screwcut throughout with a chaser held in the toolpost. For this approach, a chaser is sometimes borrowed from a die-head. However, the ISO metric and American Unified threads were designed with the limitations of lathe screwcutting taken into account, and crest radii on these forms is optional. Indeed, so also is a root radius optional, but if this is entirely omitted it calls for a cutting tool with sharp corners, prone to breakdown. However, when we use a tool with a minimum tip radius, and depth with the cross-slide and bring the pitch diameter to the desired value by top-slide advancements (top-slide parallel to bedways) we produce a very short flat root with a minute radius at each corner as shown in the diagram Fig.62.

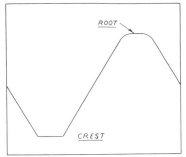

Fig. 62. Form of screw thread root and crest obtained from screwcutting by method 3.

We should note, however, that when a stud or bolt is highly stressed as in certain parts of aircraft and transport vehicles, the advisability of having thread root profiles free from abrupt changes of contour should not be disregarded: sharp thread root corners can promote stress cracks that spread and lead to failure. For this reason the writer avoids making screws or other parts of transport vehicles wherein failure of a component could lead to accident and costly legal action.

TAPERED THREADS.

When cutting tapered screw threads, the cutting sides of the tool should be symmetrically disposed about a right-angle relative to the axis of the screw, and

Fig. 63. Illustrating a method for cutting short and unimportant taper threads when a taper turning attachment is not available.

The tool is set with angle A = 90 deg.

During each cutting pass in the direction C the depth of cut is maintained by in-feeding with the cross-slide in direction F.

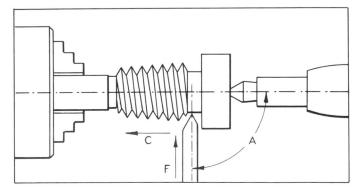

153

not at right-angles to the surface of the cone upon which the threads are to be cut. When an accurate taper is required it is necessary to use a taper turning attachment and for vee form threads to depth with the top-slide set round to half the thread angle because, of course, the cross slide cannot be moved independently of the taper motion.

For cutting short experimental or unimportant tapered threads it is sometimes possible to adopt the approach similar to that shown in Fig.63, when the thread is right-handed. Arrangements are made to cut the thread in such a way that the largest diameter is at the right, then during the cutting pass in the direction C the cross slide feed is hand advanced in direction F at the appropriate rate to maintain the depth of cut. By operating in this way, with the diameter reducing during traverse, the tendency is for the depth of cut to ease rather than to increase, and after a few passes the correct rate at which to turn the cross feed handle is readily found.

SCREWCUTTING SPEEDS

The rotational speed at which screwcutting can be carried out depends to a much greater extent than ordinary turning upon the material being threaded, the rigidity of the lathe and the means provided for stopping at the termination of the threading passes. In addition there is the question of the feasibility of engaging the half-nuts on to a fast-revolving leadscrew at the correct moment shown by reading a leadscrew indicator when this has to be used. There is little point in taking high speed threading passes if the machine has to be stopped or slowed to re-engage the half-nuts. Again, threading speeds are limited by circumstances rather than by the tool or the material being threaded. With lathes of the more customary type,

high speed threading up to a shoulder is impossible.

The rotational speed at which brass can be turned is notoriously high. A special machine such as the Hardinge Hob and Drag, for example, will thread 20 tpi in a blind bore of $1\frac{1}{4}$ in. diameter at 3000 rpm with a single-point tool. The sample witnessed had about seven turns of thread, so the time for each cutting pass was about fourteen hundredths of a second, and the thread was fully formed in about 7 passes of the tool.

The Hardinge HLV-H High-Precision Lathe illustrated in Fig. 64 is provided with auto-runout stop and a reversible dog-clutch giving repeat pickup for all thread pitches and will thread brass at up to 1000 rpm. The Hardinge lathe will also machine the hardest stainless steels with artistic ease, and to a micro-finish which frequently eliminates call for finish grinding to size. Moreover, the same hard steels can be threaded at much higher speeds and reduced overall times compared with the general run of lathes: an inch length of 7/8 in. diameter X 18 tpi can be threaded in about one minute at 350 rpm with a carbide tool. The writer attempted to thread a sample of this same hard stainless in his own lathe, but found no way in which this could be done: indeed it was hardly possible to scratch the material. It knocked the tip straight off a carbide tool, and quickly blunted an HSS tool. The secret of the Hardinge lies in its vastly greater rigidity which forces a tool to 'hold to a cut', whereas with a more lightly constructed lathe a tool is fairly easily 'pushed away' from very hard steels.

Although the Myford lathe illustrated in Fig.10, (Section 3) after modification with the special screwcutting dog-clutch control, auto-runout stop and independent tool retraction, will, for example, thread an

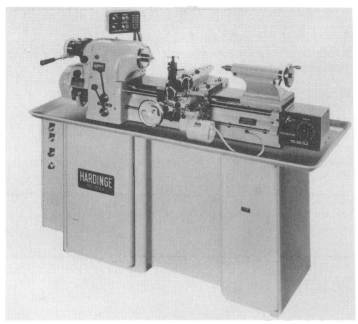

Fig. 64. The HLV-H Super Precision high-speed Toolroom Lathe. This lathe screwcuts with instant repeat pick-up for ALL threads together with automatic thread runout stop. The topslide is itself independently retractable for screwcutting.

Photo by courtesy Hardinge Machine Tools Limited, Feltham, Middlesex.

inch length of 26 tpi on 1 in. dia. brass at 500 rpm, giving cutting passes each of about three seconds duration, the controls offered by the average centre lathe simply will not allow these speeds, and broadly speaking, brass threading usually has to be carried out at speeds suitable for steels.

Free-cutting steel, as its name implies, is easily cut, and will turn and thread to a good finish at much higher speeds than will ordinary 'good commercial quality' mild steel.

As a very rough guide, when threading ordinary mild steel, initially form the threads at a speed of about one quarter of those used to turn the diameter, then reduce to about one sixth for finishing cuts - assuming high speed steel tools are being used as distinct from carbide. (Carbide does not automatically give the best results for threading in the general run of centre lathes). If more than one workpiece is to be threaded, then speeds can be increased somewhat as experience is gained.

LUBRICATION

Although for the initial stages of threading steels the usual soluble-oil-and-water serves well enough, you may notice that when the tool has lost its initial razor keenness, a depthing increment of 0.0005 inch ('half a 'thou!) (0.0127 mm.) will fail to remove any metal at all: that this is possibly due to the interposition of a water surface skin seems to be indicated by the fact that if the work is brushed clean and treated with neat cutting oil, then half 'thou. depthing increments will again take effect. For the most blemish-free results when threading commercial mild steel, tool steels and alloy steels it is therefore advisable to take light finishing passes with neat soluble oil at rotational speeds that could well be classed as "painfully slow", even though the initial thread forming can be carried out at rather more satisfying speeds.

A disadvantage of using neat cutting oil for thread finishing is that it clings to the work and fills the thread grooves, at the same time retaining minute swarf particles, all of which have to cleaned away every time it is necessary to take a wire sizing measurement. On the other hand, although a tool will generally cut after very small depthing increments in mild steel when the component is 'dry', the surface finish of the flanks will more likely be matt than bright.

SCREWCUTTING TROUBLES.

'Digging-In'

A trouble sometimes experienced during lathe screwcutting is a sudden 'digging-in' of the tool. This may produce a single 'nick', or the tool may remain in its unfavourable position and tear or 'rip' along the whole of a nearly finished thread: a circumstance that can be unfortunate if insufficient metal is left to clean up the resulting ragged helix. We may note however that digging-in is less likely when cutting internal threads, because the natural spring or 'give' in the toolshank will not allow the cutting end to'hold' to an adverse or extra heavy cut.

Digging-in can arise from a number of circumstances, although it is more likely to occur when threading steels. Here are six possible causes:

(1). The tool may be trying to remove too much metal simultaneously from both leading and trailing thread flanks, thus introducing a wedging action causing the workpiece to try to 'ride up' the tool and to tear rather than cut the flanks.

(2). The tool may be above centre and therefore unable to take a light cut. Then, when under the mistaken impression that the cut was of insufficient depth, a greater depthing is tried, the workpiece springs upwards and the tool then cuts too deeply.

(3). The workpiece may slip in the chuck or driving carrier, thus greatly increasing the amount of metal removed by the leading cutting edge of the tool.

(4) .The half-nuts may be partially disengaging under the stress from traverse loading. With the more customary Acme form of leadscrew with its sloping thread flanks, a partial disengagement of the half-nuts will introduce a lag in tool traverse, and sometimes a cutting action will cease. Then, on recommencing with an increased depth of cut, and under the mistaken impression that the previous depthing was insufficient, a cut much in excess of the anticipated amount may be taken, again resulting in flank tearing.

(5) .The cutting speed may be too high.

(6) The half-nuts may be out of alignment. The writer once encountered a lathe with non-matching half-nuts: a particularly difficult fault to trace when one does not normally blame tools for poor work. In this instance the leadscrew was of the Acme form, and as attempts at full engagement of the half-nuts resulted in severe seizure of the leadscrew, a stop was fitted to limit the depth of engagement and to thus reduce the friction. However, with the limiting stop preventing full nut engagement, the leadscrew was free to deflect within the

half-nuts, and as was afterwards found, would sometimes adopt a planetary motion within the half-nuts, rising into full engagement with the upper half-nut, then wedging therein and rolling round into full engagement with the lower half-nut, and again wedging and rolling, with disastrous effect upon the thread being cut.

The Acme thread has flanks sloping at an angle of $14\frac{1}{2}$ deg. and deflection of an Acme leadscrew therefore advances or retards a lathe carriage by tan. $14\frac{1}{2}$ deg.: 0.25862, i.e. $\frac{1}{4}$ thou. in. for each thou. in. of leadscrew deflection, and this movement was being superimposed upon the normal leadscrew lead and rendering screwcutting totally unreliable.

To check half-nuts for alignment, remove the half-nuts and leadscrew. Engage the half-nuts with the leadscrew, and use a straight-edge to check for alignment of the half-nut slideways. If necessary, re-align by filing the ways, or shears as they are sometimes called.

It is worth noting that none of the foregoing bother could have occurred had the leadscrew been of square-thread form with which the nuts would either have engaged, or if out of alignment by any significant amount, would not have entered the leadscrew threads at all, and of course, with thread flanks at right-angles to the leadscrew axis, leadscrew deflection could have no effect upon carriage position.

As a matter of fact, having experienced this profound inconvenience from the Acme form leadscrew and half-nuts, I replaced my leadscrews with threads of square form.

Manufacturers are aware of the undesirable element present in the Acme form leadscrew, but nevertheless prefer the Acme form as being cheaper and easier to produce, which seems rather strange in view of the great amount of expensive cosmetics and refinements lavished upon other details of lathe construction. There is however a hint that the Acme-form leadscrew will be replaced by a similar thread with a flank slope of 5 deg., i.e. a 10 deg. included-angle thread-form which has properties more nearly approaching those of the square form.

POOR FINISH

General note: when threading steels, thread crest burrs, until cleaned off, can often give the whole a very ragged and dull appearance. Otherwise, poor finish may result from any of the following:

(1) The tool may be below centre, and therefore cutting badly.

(2) A minute piece may have broken away from the apex of the threading tool. Examine the tip with a watchmakers' glass.

(3) If the leading flanks are bright and the trailing flanks are dull (as viewed in oblique light) a possible cause is that the trailing side of the threading tool has had insufficient metal to remove: probably from a too pedantic following of top-slide advancements for each cross slide depthing increment (top-slide set parallel to bedways). Or if working with the top-slide set at half the thread angle, the trailing cutting side of the threading tool may not have been ground or set to the precise angle needed to take advantage of the oblique setting method.

(4). It is possible to rough a thread too slowly. When a workpiece is rotated too slowly, the cutting action appears to be somewhat less smooth or uniform than at the 'higher' speeds, especially when roughing cuts are placing any significant loading on a tool. At the moment a tool commences to cut, gear backlash and general stresses are taken up, then, almost at the same instant, the actual cut slightly relieves the initial stress, and there is a momentary lag, only to be at once taken up again, with the result that the thread is formed by a kind of repeated snipping action instead of by a continuous cut. This can leave thread flanks too rough or chatter marked for subsequent light finishing cuts to take effect without undersizing.

(5). Some lathes are so designed that when cutting right-hand threads the leadscrew is under compression. That is to say that leadscrew thrust is being absorbed or resisted by the right-hand leadscrew bracket or bearer so that in effect the leadscrew in pushing, and not pulling the carriage along the lathe bed, and if the leadscrew is not of substantial diameter there may be a tendency for it to whip or buckle under stress of carriage traverse against the resistance of threading passes. To counter this undesirable design element the writer modified his leadscrew thrust by fitting a special collar at the left-hand end. Thus, when cutting right hand threads the leadscrews operate in tension and, of course, when cutting L.H. threads the leadscrews are also in tension, with the original right hand collar abutting the outer end of the R.H. leadscrew bearer.

Practical Thread Sizing Measurement

When cutting a screw thread we have to check from time to time that it is approaching the correct proportions, or, at worst, is not oversize. As already mentioned, unless a screw is for one's own use and has to fit a specific nut, then gauging with a standard soft commercial nut, or with a nut 'carefully tapped with a ground-thread tap' cannot be relied upon. In the event of dispute, one must be able to demonstrate that the thread size was checked by recognised methods that can be repeated elsewhere without reference to 'a nut'. It is for these reasons that, with the whole of the preceding eight Sections on lathe screwcutting, this, Section 9, is the one to which the writer has to repeatedly refer – simply because the formulas are not easy to commit to memory.

Sometimes, for sub-contract work, a hardened and ground-thread gauge will be supplied, but when only one or a few 'special' screws are required, such as for example, a 3 in. dia. X 16 tpi, or 75.0 mm. dia. X 1.5 mm. pitch, and well outside what we may term 'everyday sizes', then it would be unreasonable to expect to be supplied with a standard gauge which could cost £100 or so. Accordingly, any such special threads have to be checked by 'wiring': indeed, without a special gauge, and apart from the use of special 'screw-thread micrometers' there is no other way of checking before removal from a lathe.

THE WIRING FORMULAS

Although a profusion of decimal places can make the necessary formulas appear rather formidable, in practice they are quite simple, and contrary to an impression sometimes given by reference books that 'wiring' is only applicable to the checking of precision hardened and ground-thread gauges with the aid of special 'rigs' and wires sized to very close limits, the method is readily adaptable to the shop floor, and special wires are not necessary for run-of-the-mill checking.

However, before we can have a complete understanding of wire-gauge, it will be an advantage to have some knowledge of the theory upon which thread forms are based, together with the meaning of the various terms associated with threads, and these details are now offered.

DEFINITIONS OF SCREW THREAD TERMS.

The chief terms used in discussing the features of a screw thread are illustrated in Fig. 65 and are more fully defined by

159

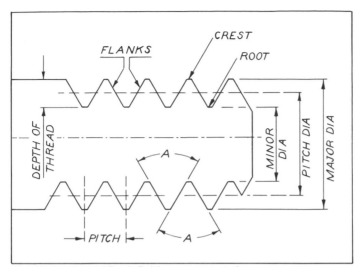

Fig. 65. Section of screw thread showing the chief terms used when discussing the various features. A = the thread angle, sometimes known as the 'included angle'.

the following 15 definitions extracted from 'Machinery's Handbook' 18th Edition by permission of the Editor:

MAJOR DIAMETER: The largest diameter of a straight screw thread. The term major diameter applies to both internal and external threads and replaces the term "outside diameter" as applied to the thread of a screw and also the term "full diameter" as applied to the thread of a nut.

MINOR DIAMETER: The smallest diameter of a straight screw thread. The term minor diameter applies to both internal and external threads and replaces the terms "core diameter" and "root diameter" as applied to the thread of a screw and also the term "inside diameter" as applied to the thread of a nut.

PITCH DIAMETER: (Simple Effective Diameter) On a straight thread, the pitch diameter is the diameter of an imaginary co-axial cylinder, the surface of which would pass through the thread profiles at such points as to make the width of the groove equal to one half of the basic pitch. On a perfect thread this occurs at the

point where the widths of the thread and groove are equal.

On a taper thread, the pitch diameter at a given position on the thread axis is the diameter of the pitch cone at that position.

NOMINAL SIZE: The nominal size is the designation which is used for the purpose of general identification. For example, the nominal size of $\frac{1}{2}$inch - 20 thread is $\frac{1}{2}$inch, but its actual size (major diameter) for Class 2A limits of size range from 0.4987 to 0.4906 inch.

ACTUAL SIZE: An actual size is a measured size.

BASIC SIZE: The basic size is the theoretical size from which the size limits are derived by the application of the allowance and tolerance.

DESIGN SIZE: The design size is that size from which the limits of size are derived by the application of tolerances. When there is no allowance the design size is the same as the basic size.

BASIC FORM OF THREAD: The basic form of thread is the theoretical profile of a thread for a length of one pitch in the axial plane. It is the form on which the

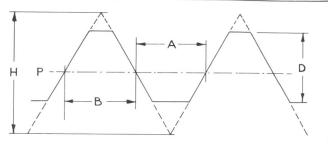

Fig. 66. Showing how the pitch line P falls at half basic triangle height H, but not necessarily at half thread depth D. In a perfect thread, space A exactly equals dimension B on the pitch line.

design forms of both the external and internal threads are based.

PITCH: The distance from a point on a screw thread to a corresponding point on the next thread measured parallel to the axis.

LEAD: The distance a screw thread advances axially in one turn. On a single-thread screw the lead and pitch are identical; on a double-thread screw, the lead is twice the pitch; on a triple-thread screw, the lead is three times the pitch, etc.

ANGLE OF THREAD: the angle included between the sides of the thread measured in an axial plane.

LEAD ANGLE: The lead angle of a straight thread is the angle made by the helix of the thread at the pitch diameter with a plane perpendicular to the axis.

CREST: The top surface joining adjacent sides or flanks of a thread.

ROOT: The bottom surface joining adjacent sides or flanks of a thread.

FLANK: The flank (or side) of a thread is the surface connecting crests and roots.

CONSTRUCTION OF THREAD

A Vee form screw thread is constructed about a fundamental triangle (Fig. 66) the height **H** of which varies with the angle of the thread. In any screw thread, the pitch line **P** always falls at exactly one half the height of the fundamental triangle, and at that height, with a perfect thread (i.e. a symmetrical thread form) space **A** will exactly equal the thickness **B** of the body of the thread. In a symmetrical thread form such as the Whitworth, the pitch line will fall at exactly one half thread depth, but with a non-symmetrical thread such as the ISO Metric and American Unified which have a root width greater than crest width, although the pitch line falls at half fundamental triangle height, it will not fall at half screw thread depth. It follows, therefore, that although a screw thread may have been machined to the correct major diameter and thread depth, it will not necessarily be correctly proportioned. This is illustrated by the three diagrams, Fig. 53, Section 8, where all three threads are drawn with the same major diameters **D**, and minor diameters, **d**, but at **A** the tool had a sharp point and at **B** a blunt point, whilst at **C**, we may assume that the tool had a minimum honed tip radius and that after depthing to the prescribed amount, the leading and trailing flanks were shaved by the appropriate top-slide adjustments, as described in Section 8, until the thickness of the body of the thread did not exceed, or was exactly equal to the width of the Vee space measured on the pitch line.

PRINCIPLE OF WIRING

That the thickness of the body of the thread does not exceed the width of the space as measured on the pitch line may be gauged by ascertaining the degree to which a rod or wire of suitable diameter will sink into a thread groove, for which purposes measurements may be made in the manner shown in the photograph, Fig. 67. When the micrometer reading agrees with or does not exceed the calculated figure and the thread is not drunken, and the thread angle and pitch are free from errors, then the basic maximum pitch diameter and satisfactory thread proportioning may be assumed.

THREE-WIRE METHOD.

Measuring the increment brought about by one wire positioned in a thread groove is a modification of what is known as "The Three-Wire Method for Effective or Pitch Diameter Measurement," a method capable of giving very accurate results

when applied in the Inspection Department. Fig. 68 illustrates the principle. Wires **W** of appropriate diameter are positioned in the thread grooves, one at one side and two at the opposite side for symmetrical abutment to the anvil of a micrometer, the spindle of which takes a reading from the single wire, so that the value of dimension **M** can be ascertained.

When a measured dimension **M** agrees with a calculated figure, a thread will be correctly proportioned with a maximum pitch diameter. If the actual value of the pitch diameter is required it can be calculated, although for thread proportioning purposes it is not necessary to know the value.

In an Inspection Department, the special steel wires used are calibrated to size within one hundred thousandth of an inch (254 millionths of a mm.) and with these wires held in a special rig, pitch diameter checking can be made to an accuracy of a tenth of a thousandth of an inch (0.00254 mm.). However, it is hardly necessary to aim at such a high degree of accuracy when checking work in the lathe because the lightest cut a lathe is capable of taking from a thread flank will probably not be less than half a thou. inch (0.0127 mm.) In other words, if an attempt is made to remove less, a tool will often cut rather more than that amount or will remove nothing at all. Accordingly and with but slight accuracy loss we may use twist drill shanks or domestic sewing needles instead of the more accurate steel wires, although a careful measuring of the actual diameters of drill shanks or needles

Fig. 67. Checking a thread for sizing by ascertaining to what degree a wire — twist drill shank or needle — sinks into a thread groove. When an actual reading agrees with a calculated reading, the thread will be correctiy proportioned.

is advisable, even for lathe work, because "wire" diameter errors become magnified in the formulas. Measurement to a tenth thou. inch or to 0.002 mm. should, however, give every satisfaction.

WIRE DIAMETER AND THREAD DEPTH.

The ideal or "best" diameter for any wire is that which will contact the thread flanks at the point of intersection of the pitch line, but in practice there is considerable latitude. For example, for checking Whitworth threads the smallest satisfactory wire = 0.54 X pitch (or 0.54/T.P.I.) and the largest diameter = 0.76 X pitch (or 0.76/T.P.I). In this respect it is worth noting that the depth of a standard Whitworth screw thread is 0.64/tpi and that 0.64 falls so comfortably between 0.54 and 0.76 as to permit of suitable wire diameters being taken as approximately equal to the depth of the thread: figures easily remembered or

calculated from the formulas given and equally well applicable to ISO Metric, American Unified, and Acme threads.

Examples of 3-wire checking will lead to an understanding of the simplified single-wire method and the associated formulas for the latter. Let us therefore find what value the **M** reading should have over three wires when a 30.0 mm. dia. X 2.5 mm. pitch ISO Metric thread is correctly proportioned.

The 3-wire formula used for 60 deg. Vee threads reads:

$$M = D-(1.5155 \times P) \text{ plus } (3 \times W)$$

where D = major screw diameter
(design size)
P = pitch of thread.
W = diameter of wires.

As previously mentioned, wire diameters can approximately equal thread depth. The depth of thread for 2.5 mm. pitch ISO Metric = 2.5 X 0.6134 = 1.5335 mm. Accordingly we may select three 1.5 m.. twist drills and assume for our purposes here that the measured shank diameters

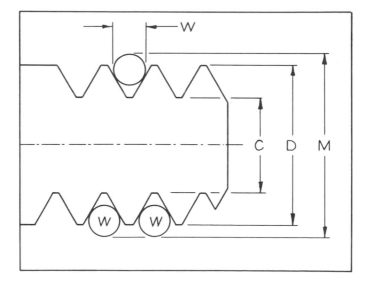

Fig. 68. Principle of the three-wire pitch diameter checking method. Note: Measurement M over wires W ignores both major diameter D and minor diameter C. Nevertheless, this wire checking would easily reveal the faults in a screw at A or B Fig. 53 (Section 8).

have a value of 1.492 mm., each being 0.008 mm. under nominal size.

Substituting the known figures we have:

Example (1)

$M = 30 - (1.5155 \times 2.5)$ plus (3×1.492)

$M = 30 - 3.78875$ plus 4.476

$M = 26.21125$ plus 4.476

$M = 30.68725$ mm.

An ordinary vernier reading metric micrometer will show only three decimal places, accordingly we may take **M** as 30.687 mm.

Let us now suppose that we have to gauge the same metric thread entirely by inch measure. The major screw diameter will equal 30X 0.03937 = 1.1811 inch, and with **M**, **D** and **W** in inches the 3-wire formula (with conversions written in where necessary) reads:

$M = D - (0.0596 \times P(mm))$ plus $(3 \times W)$

Again selecting the diameter of the wires **W** as being approximately equal to thread depth, we may note that the depth of an ISO Metric screw thread in inches = P (mm) × 0.0241, so the depth of the 2.5 mm. pitch thread = 0.06025 inch, say No. 53 drills with a nominal shank diameter of 0.0595 inch. Assuming the shanks are undersized by 0.0003 inch, we will use a figure of 0.0592 inch for the value of W. Substituting the known figures we have:

Example (2)

$M = 1.1811 - (0.0596 \times 2.5)$ plus
$$(3 \times 0.0592)$$

$M = 1.1811 - 0.1490$ plus 0.1776

$M = 1.0321$ plus 0.1776

$M = 1.2097$ inch.

Accordingly we may assume that when the actual reading over the drills is equal to or does not exceed 1.2097 inch, the thread spaces and body thickness at the pitch line will be equal.

THREE WIRE FORMULA FOR 55 deg. SCREWS.

The 55 deg. thread angle of the Whitworth form requires the use of different figures in the formula because the more acute thread angle will not allow a wire to enter quite so deeply as it would in the corresponding 60 deg. thread Vee.

The three-wire formula for 55 deg. threads sized by inch measure reads:

$M = D - (1.6008 \times P)$ plus $(3.1657 \times W)$

but to avoid the inconvenience of having to calculate or look up the pitch figure when threads are given in terms of threads per inch, we may use 1.6008/ T.P.I. in place of 1.6008 x P and the formula will read:

$M = D - \left(\dfrac{1.6008}{T.P.I.} \right)$ plus $(3.1657 \times W)$

Let us now calculate the M reading over the three wires for a 10 T.P.I. thread with a major diameter of 2 inches. Taking **W** as equal to thread depth: 0.64/T.P.I., **W** may equal 0.064 inch, say No. 52 drill with a nominal diameter 0.0635 inch, and let the measured shank diameters equal 0.0632 inch. Substituting the known figures we have:

Example (3)

$M = 2.0 - 0.16008$ plus 0.20007224

$M = 2.03999224$, say

$M = 2.0400$ inch.

WHITWORTH BY METRIC MEASURE.

When a Whitworth thread is given in terms of tpi and it is required to machine and gauge by millimetre measure, the following formula with built-in conversions may be used with **M**, **D** and **W** in millimetres:

$$M = D - \left(\frac{40.66}{T.P.I.}\right) \text{ plus } (3.1657 \times W)$$

So that we may compare the metric **M** figure with the inch **M** figure in Example (3) let us calculate the **M** reading for a thread of 10 T.P.I. on a diameter of 50.8 mm. using for **W** the same No. 52 drills with measured shank diameters converted into millimetres. **W** will therefore equal 0.0632 × 25.4 = 1.60528 mm., and substituting the known figures, we have:

$$M = 50.8 - \left(\frac{40.660}{10}\right) \text{ plus }$$
$$(3.1657 \times 1.60528)$$

$$M = 50.8 - 4.066 \text{ plus } 5.081834896$$

$$M = 51.816 \text{ mm.}$$

Converting to inches for comparison with the **M** reading in Example (3) we have 51.816 × 0.03937 = 2.03999592 inch, a discrepancy of plus 3.68 millionths of an inch, or 93.472 millionths of a millimetre.

ONE WIRE THREAD CHECKING

From a practical point of view we may note that in taking an **M** reading over three wires, the measured or actual major diameter of the screw is ignored, and the readings are not therefore affected by thread crest burrs raised during machining. However, taking three wire measurements in a lathe can be awkward, although it is worth noting that in certain instances it is sometimes possible to use a vice to grip three needles temporarily in position on a finished or nearly finished thread, and then to force the sharp needle ends into a piece of cork, for example, thus facilitating repeated use. Occasionally wires or similar can be held in position with thick grease. When three wires are to be used, all should be of identical diameter to very close limits, as measured with a good micrometer.

However, if we examine the question in a different way, we can devise a more straightforward approach to thread

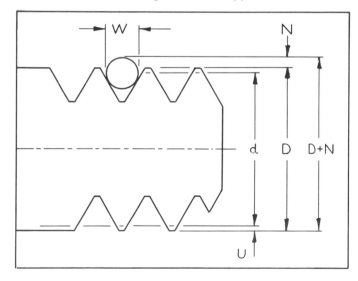

Fig. 69. Illustrating the principle of a 'one-wire' method for checking that a thread is correctly proportioned.

This method is particularly suitable for checking a thread during lathe screwcutting. Please also see text.

165

proportion checking by using only one wire, drill-shank, or needle.

Referring to Fig. 69, if we can calculate the magnitude of **N**, the amount by which one wire of known size will protrude above the crests of a perfect screw thread when the major diameter **D** equals the design diameter, we would need only one calculation for each pitch, regardless of the design diameter upon which that pitch is cut: at least for the general run of engineering requirements.

The formula for ascertaining the value of the increment **N** by millimetres for a 60 deg. metric screw thread with **N**, **P** and **W** in millimetres reads:

$$N = \frac{(3 \times W) - (1.5155 \times P)}{2}$$

Taking our previous Example (1) we may now rephrase and simplify the question: "Find the value of **N** for an ISO Metric screw thread of 2.5 mm. pitch":
(Let W = 1.492 mm., as in Example (1))

Substituting the known figures we have:
Example (5)

$$N = \frac{(3 \times 1.492) - (3.78875)}{2}$$

$$N = \frac{4.476 - 3.78875}{2}$$

$$N = \frac{0.68725}{2}$$

$$N = 0.343625 \text{ mm.}, \quad \text{say } 0.344 \text{ mm.}$$

The advantage of this approach lies in the fact that the **N** value of 0.344 mm. holds for all 2.5 mm. pitch ISO Metric or other 60 deg. screw threads regardless of the diameter upon which that pitch is cut, and it is not at all difficult to take the **D** plus **N** reading whilst the work is mounted in the lathe, although, of course, crest machining burrs must be removed by a light filing before measuring, and, as with the three-

wire method, it is also advisable to wire-brush the thread grooves.

On Example (5), had the 2.5 mm. pitch been cut on a major diameter of 30.0 mm., measurement **D** plus **N** would read 30.344 mm., when the body of the thread had been machined to the correct proportions. Had the measured major diameter been machined undersize by, say, 0.05 mm. (approx. 0.002 inch) through normal allowances, then, when the thread is properly proportioned the micrometer would read half that amount (0.025 mm.) less than the calculated **N** reading:

$$(D \text{ plus } N) - \left(\frac{D - d}{2}\right)$$

where $D=$ the design diameter, and d the measured diameter, accordingly:

$$30.344 - \left(\frac{30.0 - 29.95}{2}\right)$$

$$30.344 - \frac{0.05}{2}$$

$$30.344 - 0.025$$

$$= 30.319 \text{ mm. to show satisfactory proportioning.}$$

The practical operator will, of course, interpret the foregoing major diameter adjustment formula as "Deduct from **D** plus **N** one-half the difference between the major design diameter and the actual turned diameter; amount **U** Fig 69, or, in workshop language: "knock off half the difference."

We may now also note that twice the calculated **N** figure gives the increment for three wires, for example 2 X 0.343625 = 0.687250 mm. giving the figures we would obtain by deducting the 30 mm. major diameter from the **M** reading calculated in Example (1). Accordingly, if the **N** figure for any single pitch is known, then 2 X **N** plus the major (design) diameter will give a figure for the **M** reading over three wires as a final check after a series of more readily made single wire measurements when a screw is approaching completion.

Although the allowances on wire diameters are high enough to permit of any one size being used for more than one thread pitch, in the workshop we are really only concerned with one pitch at a time: it therefore seems a more straightforward approach to adhere to the fact that wire diameter may approximately equal the depth of the thread to be checked, and to choose a wire, drill shank or needle for each pitch, at least in the early stages. Once a chosen wire has been measured as nearly as possible to a tenth of a thou-inch, or to 0.002 mm. and the **N** calculation made for the pitch to be cut, the wire may be slipped into a card folder together with a note of the pitch, or threads/inch and the **N** reading, ready for repeated use, for example.

(1) No. 54 drill shank — measured dia.
0.0547 inch.

12 T.P.I. 55 deg. Vee thread N − 0.0199 in.

12 T.P.I. 60 deg. Unified   N = 0.0189 in.

(2) Needle. Measured dia. 0.0183 inch.

ISO Metric 0.75 mm. pitch   N = 0.005 in.

It is interesting to note that if a wire or needle of too small a diameter is chosen for any given pitch the fact will become evident from an inability to obtain a positive value from the subtraction required by the formulas.

PITCH DIAMETER CALCULATIONS.

As will have been seen, although the three-wire measurement is referred to as the wire method for measuring effective or pitch diameter of a screw thread, it does not give the actual or directly measured value: to find this from the wire method we need an actual **M** reading over the three wires, and another formula.

The effective diameter can be measured directly only with special micrometers having conical ended spindles and Vee shaped anvils. A considerable number of micrometers is required to cover a range of screws, and the number is still further increased by the necessity for having whole sets to suit various thread angles, therefore screw thread micrometers are not likely to be found in any but the most lavishly equipped workshops. On the other hand, for lathe screwcutting it is not often that a knowledge of the actual pitch diameter is of any help: pitch diameter will be at a maximum value when the measured **M** or **N** reading agrees with a calculated reading, and if it is necessary to increase the pitch diameter, as perhaps for a tap, this can be done by machining until the measured **M** reading is the desired amount larger than the calculated amount. In this respect it should be noted that if **N** readings are being taken over only one wire, they are radius readings: 0.0005 inch (0.0127 mm.) above the calculated figure will give an effective pitch diameter increase of 0.001 inch (0.0254 mm.).

If it is desired to ascertain the actual or "measured" pitch diameter of a screw being cut, this can be done after an **M** reading has been taken over three wires, then, when **M**, **W** and pitch (or T.P.I) are known, the actual pitch diameter for 60 deg. Vee threads can be found from:

$$E = M \text{ plus } (0.86603 \times P) - (3 \times W)$$

or, if the thread is in terms of tpi:

$$E = M \text{ plus } \left(\frac{0.86603}{TPI}\right) - (3 \times W)$$

Let us calculate the pitch diameter for Example (1) (page 164) when **M** over three wires reads 30.687 mm. for 2.5 mm. pitch ISO Metric 30.0 mm. dia., and wire diameter = 1.492 mm.

$$E = 30.687 \text{ plus } (0.86603 \times 2.5)$$
$$- (3 \times 1.492)$$

$$E = 30.687 \text{ plus } 2.165075 - 4.476$$

$$E = 28.376075 \text{ mm.}$$

Similarly the pitch diameter for 55 deg. threads can be found from:

$$E = M \text{ plus } \left(\frac{0.9605}{TPI}\right) - (3.1657 \times W)$$

THE ACME FORM THREAD.

In the past some concern was felt that because of the steeper slope of the Acme thread flanks the measuring pressure on a wire would force it too deeply into the thread groove to give a correct reading. However, this point seems recently to have been resolved because 'Machinery's Screw Thread Book' now gives a formula for **M** readings over three wires:

$$M = D - (2.433566 \times P) \text{ plus}$$
$$(4.9939292 \times W)$$

As with the other 3-wire **M** formulas I have re-arranged it for calculating the **N** increment values for one wire so that the formulas can be used on any diameter. For sizing by inches or millimetres, four formulas are required, with built in conversions where necessary: these will be found at the end of the following summary of formulas for ascertaining **N** values under various language conditions.

SUMMARY OF *N* VALUE FORMULAS

(1).

60 deg. Vee thread, sized by millimetres.
    Pitch in millimetres.
    (N, W and P in mm.)

$$N = \frac{(3 \times W) - (1.5155 \times P)}{2}$$

(2).

60 deg. Vee thread, sized by inches.
    Pitch in millimetres.
    (N and W in inches)

$$N = \frac{(3 \times W) - (0.0596 \times P)}{2}$$

(3).

60 deg. Vee thread, sized by inches.
    Pitch in terms of T.P.I.
    (N and W in inches)

$$N = \frac{(3 \times W) - \left(\frac{1.5155}{T.P.I.}\right)}{2}$$

(4).

60 deg. Vee thread, sized by millimetres.
    Pitch in terms of T.P.I.
    (N and W in mm.)

$$N = \frac{(3 \times W) - \left(\frac{38.4937}{T.P.I.}\right)}{2}$$

(5).

55 deg. Vee thread (Whitworth & B.S.F.), sized by inches.
    Pitch in terms of T.P.I.
    (N and W in inches)

$$N = \frac{(3.1657 \times W) - \left(\frac{1.6008}{T.P.I.}\right)}{2}$$

(6).

55 deg. Vee thread, sized by millimetres.
    Pitch in terms of T.P.I.
    (N and W in mm.)

$$N = \frac{(3.1657 \times W) - \left(\frac{40.66032}{T.P.I.}\right)}{2}$$

(7).

ACME form thread, sized by inches.
Pitch in terms of T.P.I.
(N and W in inches)

$$N = \frac{(4.994 \times W) - \left(\frac{2.433566}{T.P.I.}\right)}{2}$$

(8).

ACME form thread, sized by millimetres.
Pitch in terms of T.P.I.
(N and W in millimetres)

$$N = \frac{(4.994 \times W) - \left(\frac{61.81257}{T.P.I.}\right)}{2}$$

(9).

ACME form thread, sized by millimetres.
Pitch in millimetres.
(N, W and P in millimetres)

$$N = \frac{(4.994 \times W) - (2.433566 \times P)}{2}$$

(10).

ACME form thread, sized by inches,
pitch in millimetres.
(N and W in inches)

$$N = \frac{(4.994 \times W) - (0.0958 \times P)}{2}$$

The necessary built-in conversion figures
were obtained as follows:

In formula (2) 0.0596 is derived from

1.5155 × 0.03937 (= 0.059665235) to
convert to inches.

In formula (4):

38.4937 = 1.5155 × 25.4 to convert to
millimetres.

In formula (6):

40.66032 = 1.6008 × 25.4 to convert to
millimetres.

In the Acme *N* formulas I have taken the
liberty of rounding off 4.9939292 to 4.994
for the *W* multiplier, and in formula (8)
61.81257 is derived from

2.433566 × 25.4
to convert to mm.

In formula (10) 0.0958 is derived from
2.433566 × 0.03937
to convert to inch measure.

HELIX ANGLE OF A SCREW THREAD

The helix angle of a thread is also
known as the lead angle, and is illustrated
in Fig. 70.

All the foregoing M, **E** and **N** formulas
are suitable for the general run of screw
threads having helix angles insufficiently
pronounced to affect the wire position to
any degree capable of being corrected
during lathe screwcutting, the errors
seldom reaching 0.0001 inch (0.00254
mm.)

The effect of a pronounced helix angle
on a wire used for taking pitch diameter
readings is to prevent the wire entering
the thread groove so deeply as it would
with a less severe helix angle, unless, pre-
sumably, the wire is of a small gauge that
can be wound around the thread groove in
the form of a helix. So if a helix angle is
not taken into account for threads of high
lead, and the thread is sized to an
ordinary **M** or **N** reading, the body of the
thread would be undesirably thinned
before the measured M reading agreed
with the calculated M reading. Those
interested in this aspect will find it fully
dealt with in 'Machinery's Screw Thread
Book,' Section J.

The helix angle of a thread taken from a
plane at right-angles to the axis may be
found from a formula which first gives the
tangent of the angle. The angle itself is

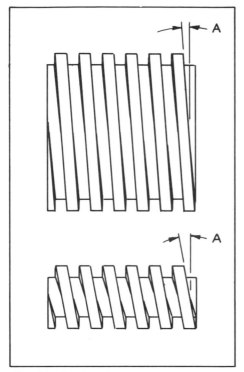

Fig. 70. As the diameter of a thread of any given pitch decreases, so the helix angle A increases. The helix angle is also known as the 'lead angle'.

then taken from a trigonometrical table of tangent values for degrees.

$$\frac{\text{Tangent of}}{\text{Helix angle}} = \frac{\text{Lead of screw}}{\text{Pi} \times \text{Pitch diameter of screw}}$$

The diagram. Fig. 70 shows how the helix angle increases with a decrease in the diameter of a thread of given pitch.

As an example, let us find the helix angle for an ISO Metric thread of 3.0 mm. pitch on a nominal 30.0 mm. major diameter:

Lead of screw = 3.0 mm.

Pitch dia. = Major dia. − 0.65 × P
= 28.05 mm.

accordingly:

$$\frac{\text{Tangent of}}{\text{Helix angle}} = \frac{3.0}{3.1416 \times 28.05}$$

$$= \frac{3.0}{88.12188}$$

Tangent of angle = 0.03404

The table of tan. figures shows that 2 deg. has a value of 0.03492.

For the helix angle of a thread to show a more significant number of degrees it is necessary to turn to threads in the multiple-start range on fairly small diameters. For example, let us find the helix angle for a 6-start thread of 4.0 mm. pitch on a major diameter of 25.0mm.

The lead of a multiple-start thread equals the number of starts multiplied by the pitch, accordingly:

Lead of screw = 6 × 4 = 24.0 mm.

Pitch diameter = 22.4 mm., thus:

$$\frac{\text{Tangent of}}{\text{Helix angle}} = \frac{24}{3.1416 \times 22.4}$$

$$= \frac{24}{70.37184}$$

$$= 0.3410455$$

A table of tan. figures shows that a value of 0.34107 corresponds to an angle of 18 deg. 50′.

NUT SIZING.

(All internal threads are termed 'nut threads').

As explained in more detail in Section 8, when cutting an internal thread with a single-point lathe tool it is customary to gauge the progress with either the corresponding lathe cut screw, or with a special thread gauge, or to finish size with a tap provided or made in the workshop. The more usual requirement is to obtain a pleasing fit without bind or shake. When a nut or nuts only are required and there are no instructions to make corresponding screws, and the number required would not warrant purchase of a screw gauge or tap for finish sizing, then one has no option but to first cut a screw for use as a gauge, sizing the screw by wiring. Ideally such a gauge should be hardened.

THREAD CLASSES.

Although we cannot here go into lengthy detail on the various thread classes, one example will show the kind of tolerances allowed. But let us first set out the tolerance classes for the ISO Metric thread as shown in 'Machinery's Screw Thread Book':

Type of Fit.	NUT Tolerance class	BOLT Tolerance class
CLOSE	5H	4h
MEDIUM	6H	6g
FREE	7H	8g

The type of fit is, of course, self explanatory, but from the example to follow for the medium class nut and bolt we may infer that except in the smaller diameters and pitches, threads carefully cut in the lathe will tend to fall much nearer to the close fit range than to the medium.

So that the figures in the following example shall have a ready meaning without further calculations by the reader I have set them out from a turner's point of view by giving, for the screw, basic or nominal size minus so much to give maximum permitted size, and minus so much again to show tolerance. The nut allowances are best revealed by the written word.

ISO METRIC SCREW THREAD MEDIUM FIT CLASS 6g.

Nominal diameter 30.0 mm.
Pitch   3.5 mm.

Major diameter
    Nominal   Turn o/d to nominal dia.
    30.0 mm.   minus 0.053 mm. (0.0021 inch)
(1.1811 inch) minus 0.478 mm. (0.0188 inch)
              Tolerance 0.425 mm. (0.0167 inch)

Basic Pitch dia.
    Maximum
27.727 mm.   minus 0.053 mm. (0.0021 inch)
(1.0916 inch) minus 0.265 mm (0.0104 inch)
              Tolerance 0.212 mm. (0.0083 inch)

("Drop" the calculated M reading over three wires by, say, 0.05mm. to 0.25 mm. (0.002 to 0.010 inch))

Minor diameter
(From nominal major dia.
minus 2 × thread depth)
25.7062 mm minus 0.053 mm. (0.0021 inch)
(1.0120 inch) minus 0.517 mm. (0.0203 inch)
              Tolerance 0.468 mm. (0.0182 inch)

From the point of view of thread depthing, the basic depth from major nominal diameter is 2.1469 mm. (0.08436 inch). If the minor diameter is taken to the full low limit, then the thread depth from full major diameter minus 0.053 mm. is extended by approximately 0.203 mm. (0.008 inch).

For the corresponding NUT: Medium fit, Class 6H:

The MINIMUM major diameter (the diameter to the thread roots) is the same as the nominal major diameter of the screw, and no tolerance is given for extending the thread depth to increase the major nut diameter above the minimum; so for lathe threading we may depth until a properly sized screw will enter the nut - if we wish to work in that way.

The pitch diameter of the nut rises to 0.280 mm. (0.0110 inc) above the basic maximum for the screw, and with a minor diameter of:

Basic Major dia. minus (Pitch X 1.0825) = 26.211 mm. minimum. There is a tolerance of 0.560 mm. (0.0220 inch) showing that the minimum bore diameter may be increased by that amount.

To the apprentice beginning to associate ordinary quality lathe turning with limits of plus or minus 0.025 mm. (0.001 inch) or less on plain diameters and bores, the high permitted tolerances on the sample screw and nut will no doubt help to inspire confidence when screwcutting is called for, especially with the added knowledge that for the vast majority of lathe screwcutting the only requirement is that a screw shall hold well up to, but be sized within its basic or nominal dimensions, and that the corresponding nut shall offer a comfortable fit to the screw.

At this juncture it will be appropriate to

from an 8 T.P.I. leadscrew with gearing in the ratio 30:38, the pitch actually given (assuming a perfect leadscrew) will be 2.5065787 mm., and strictly speaking this figure should be multiplied by 1.5155 in the 3 or 1-wire formula.

THREADS DESIGNATED BY CLASS.

In those instances where a drawing does refer to a thread by class, then of course, the limit figures have to be taken from the appropriate reference book or chart. In this respect we may note that a complete designation of a screw thread gives details of the thread system, the size and pitch of the thread, and the tolerances applicable to the thread. An example is given below (from 'Machinery's Screw Thread Book'):

Designation for an internal thread (nut): M6 × 1-6H

Designation for an external thread (bolt): M8 × 1.25-g

Thread system symbol for ISO Metric ⎤
           (general purpose)

Nominal diameter of thread in mm. ⎦

Pitch in mm. ———

Tolerance class designation. ———

mention that despite there being carefully worked tolerances on diameters, thread depths, and so on, no one has seen fit to treat pitch and thread angle similarly.

Errors in angle and pitch would, of course, affect wire readings, but with jig-ground tools one can reasonably assume that thread angles will be correct.

Regarding pitch, this also may be assumed correct when lathe gearing is exact and not an approximation. If, however, one is cutting say 2.5 mm. pitch

It is, however, worth noting that a major screw diameter often forms an extension to a shaft or turned step, and the shaft diameter limits are frequently shown as applicable also to the major screw diameter on the assumption that screw-blank and shaft will be finish turned at one pass. A subsequent removal of any thread crest burrs will then ensure that whatever has to fit the plain portion will not be prevented from passing over the thread.

Appendix 1   LIST OF TABLES